My Life in the
North Woods

Books by Robert Smith

Fiction

HOTEL ON THE LAKE

THE HUMAN IMAGE

ONE WINTER IN BOSTON

THE ORDEAL OF MR. BLAIR

Nonfiction

THE MASSACHUSETTS COLONY

THE INFAMOUS BOSTON MASSACRE

BABE RUTH'S AMERICA

MACARTHUR IN KOREA

MY LIFE IN THE NORTH WOODS

Sports

BASEBALL

BASEBALL IN AMERICA

HEROES OF BASEBALL

PIONEERS OF BASEBALL

ILLUSTRATED HISTORY OF BASEBALL

PRO FOOTBALL

GREAT TEAMS OF PRO FOOTBALL

ILLUSTRATED HISTORY OF PRO FOOTBALL

HOW THE PROS PLAY FOOTBALL

My Life in the North Woods

by Robert Smith

The Atlantic Monthly Press

BOSTON / NEW YORK

FIRST EDITION

LIBRARY OF CONGRESS CATALOGING-IN-PUBLICATION DATA

Smith, Robert, 1905–
  My life in the North Woods.

  1. Smith, Robert, 1905–   . 2. United States—
Biography. 3. Maine—Biography. 4. Outdoor life—
Maine. I. Title.
CT275.S5493A3   1986   974.1′04′0924   86-10904
ISBN 0-87113-074-2

BP
Published simultaneously in Canada
PRINTED IN THE UNITED STATES OF AMERICA

To Sheila and Barbara,
with gratitude for their timely encouragement
and with love

THIS book, begun at the persistent urging of my late dear friend, Ilona von Karolyi, would not have left the starting gate but for the earnest encouragement of Charles Everitt. It was completed with the prompting and commonsense advice of Ted Weeks.

I am grateful for the tireless efforts of Harold Matson on my behalf, for the words of praise I received from Sue Edwards, and for the practically flawless typing and the flow of kind words from Grace Wherry.

My Life in the
North Woods

❧ 1 ❧

THIS was a half-century ago, during what some people still recall as the Great Depression, and times were hard for everybody, but I used to think that they were especially hard for me. For I should have been in college instead of grubbing for small wages in the woods two hundred miles from home. And it had never been a project of mine to be there at that forlorn railroad station, late on a cold December afternoon, to pick up a shipment of sled bolts and carry them five miles to the lumber camp where I was earning not much more than my keep.

A package of sled bolts, in that day, weighed 75 pounds. When I had picked them up in the freight station, tucked them into my knapsack and hoisted them on my back, they raised my weight to just over 200. The lake ice was barely 2 inches thick, strong enough to sustain a man, or at least to sustain me on the way over easily enough, who weighed 130 unburdened. On my journey back, however, the ice seemed not nearly so secure. When I had gone half a mile from shore, on the three-mile trip back to camp, the ice began to sag beneath me like the floor of a loft, seeming to rise and fall ever so gently under my weight.

It had grown dark. A single gleam of light in the black distant border of woods marked the lumber camp. I dared not stop, for the frail ice seemed like to snap if I laid too long a strain on one spot. And it was already too late to turn back for shore, for the ice surely was no safer behind me than it was ahead. This at least was the argument I gave myself. But I suspect too I was prompted by my fear that to go back and own myself uneasy about crossing the solid ice would open me to secret ridicule. Men indeed had been moving out over the ice and back all day. I wonder, however, if any had cut straight across the lake at its widest spot, as I was doing, or if they had felt the ice sink and rise up again beneath them, with an occasional faint grinding noise as if the whole thousand-acre sheet might suddenly rend itself down the middle and let the black water boil up.

There was no doubt in my mind that, should the ice crack beneath me, I would plunge like a dropped anchor straight to the bitter depths, unable to wrench myself free from the straps that cut into my flesh. I had pulled the pack snug to bring the weight high on my shoulders, as I had been taught to do. (Now I know that a pack often rests easier if part of its weight bears on the hips.) And to keep from being borne backwards, I leaned against the weight and plunged ahead, with my eye on that tiny far-off light. Fear too urged me along, for with each rhythmic sagging of the ice my poor stomach sank as if the world were falling away from beneath me forever. Desperately I leaned into the weight of the packaged bolts, yearning toward that light and whispering frantic curses (or were they prayers) — Oh, God, I told myself, if it just will not crack on this step! Yet all the while my chest was tightened by the conviction that soon — Oh, Jesus Christ! — the snow-crusted ice would snap like a soda cracker, leaving me time for just one final desperate cry before I was utterly lost to sight. There were tears on my cheeks now from the cold and the searing pain. Yet I dared not halt and unhitch my pack, as I would on land, to let my howling muscles relax. I had to plunge on and on and on, whimpering finally like a child who is desperately late, with the light seeming to fall away from me as I could make out more and more of the ghostly ice that stretched between.

Long before I approached close enough to shore to descry clearly where it began I had started to cry aloud to myself, as if I were calling on someone to come help me. The light now at last did begin to grow nearer. It took shape at first, a sort of yellow oblong. Then I could see that it flowed from two small windows. Oh, Jesus, Jesus, Jesus, I panted now, leaning ever more obliquely toward the goal. Oh, Christ, let me reach the shore! The ice rose and fell in a wave as wide as the whole lake. Every few moments it would creak reluctantly the way boards do when too much weight is laid upon them. I told myself out loud now that if I could just last a few more yards, perhaps a rod or two, I would reach a spot where I would plunge at least into water that would not cover me. Or did the shore fall off sharply here, so near the mountain? Oh, Jesus, yes! A ten-foot pickpole here would not touch bottom!

Now and then, when I would give over muttering to myself, I could hear my own breath rasping in my throat, the noise of a hunted animal being run to earth. The black woods seemed to back away as I moved toward them. Then, without warning, I felt brush beneath my feet, small twigs like fingers that reached up through the ice. I was within yards of solid ground. The light from the windows of the office shack made two misshapen patterns on the yard, where logs lay one over the other waiting to be sawed into stove lengths and split. Instead of moving along the edge of the ice to where the dock reached out, I clawed my way through a brief run of alders and clambered up the bank to where I knew the grass began. I should have unhitched my pack here. But I had aimed so desperately for that light — the light given out by the single oil lamp in the office camp — that I would not be daunted now, despite the screams of my shoulder muscles. With my face contorted, and my breath searing my throat, I stumbled to the steps, had to take hold of the upper step with one hand to keep from toppling, then pushed the door open and plunged into the cabin.

There were two men sitting there. But I saw only Paul Cyr, a Frenchman bigger than a bear, who looked up from a magazine he had been holding and stared at me with open mouth. He had a full black beard and was missing his upper front teeth. Agile as a child, he hopped to his feet in time to catch hold of the pack I wore and slip it deftly right off my back. With that I sprawled headlong to the floor, leaving Paul with all the weight in his hands.

"Jeesa Chri'!" Paul yelled, then took to laughing. Harold, the scaler, who sat in the other chair, just moved his feet to get them out of my way. "Sunnama beetch!" Paul cried. "You carry *wan* sunnamabeetching load! Hey! You make good man carry booze! What say? We go next week to CanaDA?"

I won't say this accolade rewarded all my terror and pain. But I was able to flop into a chair and stick my legs out toward the stove and rub my tortured arms and shoulders without shame. Nor did I fail to feel some flush of pride when Paul offered my pack to Harold to heft:

"You tote a load like that, scaler? Sunnama beetch!" Harold took hold of it and grunted.

6

"Shit a god*damn!*" he exclaimed. "What the hell you got in here? Window weights?"

"Sled bolts," I said.

"That bastard weighs eighty pounds!" Harold added respectfully, giving the pack a slight hitch before setting it on the floor.

"Seventy-five," said I.

"Eighty with the wrappings and the pack," Harold muttered. "Too fuckin bad he couldn't hire a horse."

But "he," meaning Wallace, who ran the camp, already had horses enough, as Harold well knew. Besides, a horse would have surely gone through the ice at this season, as I nearly did.

Almost at that instant there was a clumping on the steps and we heard the steady blowing sound that Wallace's gas lantern gave out. Wallace, who must have caught sight of me slogging up across the yard, rattled the door a little as he opened it, although there was really no need to give warning. He came in squinting across the light of his own lantern.

"Paul," he said. To me and the scaler he nodded mildly. He took a seat on a bench that ran alongside one of the bunks and set his lantern tenderly at his feet. "These things give a lot of heat," he said. Then he sat silent and slumped a little, with his hands comfortably together between his knees, like a man in church. The whaaaaaaaw of the gas lamp, as it blew its steady breath through the glowing mantle, seemed to hold everyone's ear. Wallace looked idly about the cabin and let his glance rest for a moment or two on my pack, which lay now at the foot of one bunk.

"You fetch the papers?" he said, without lifting his eyes to me. I had picked up the letters at the post office but I had told Ray, the assistant postmaster, to hang on to the pile of rolled-up Boston *Post*s. "I had all I could lug without them," I said.

Wallace looked at me now, his mouth tight and his eyes glinting.

"Well, that woman's all alone now without a thing to read!" he declared, as if he would rivet me with guilt. "How much extry trouble would *they* have given you?"

"I would have had to carry them loose in my hand," I said.

"Well." He glowered now at the pack. "You brought the freight at least?" He gave this question a sarcastic lilt and devoted a bitter sidelong glance to my worthless self. I merely nodded and after three seconds of silence, Wallace got up and took hold of the pack. It did not come off the floor quite so readily as he had thought. He had to use both hands to lift out the package.

"Them sled bolts," he said. And no one was prepared to contradict him. He clutched the package in both hands and sat down again, resting the package beside him. Tim, Wallace's brother, who acted as boss in the woods, came in then, pausing for a while in the door to draw his mouth down in a grimace of surprise to see big Paul Cyr among us. Tim was exactly the same size as Wallace, a short-legged man who, like his brother, made himself look smaller still by walking in bent-kneed fashion, as if he were still trying to keep his footing on a floating log. Tim was older and had never married. His face lacked the look of studied innocence that Wallace wore. It was red and longer in the chin and in the upper lip than his brother's was, giving him the appearance almost of a cartoon Irishman.

"Well," said he, "the gang's all here!"

"Ummm," said Wallace.

Tim nodded several times at Paul.

"Find something good to read, Paul?"

It was not a question that sought an answer, for Tim turned away almost at once to find a place to rest his own rear end. Paul just waved his magazine as if he would either throw it away or keep it and no matter.

"Fuckin magazine," he muttered.

Everyone was silent then and the blowing of the gas lamp filled the room once more. Wallace held his hands over the light and rubbed them briefly together.

"These things give off a lot of heat," he said.

"Ummm," said Tim.

There was another long silence then. Paul, tipping his magazine to get better light on it, squinted hard at the page. Harold and I looked at each other and Harold winked. We had often sat this way before, sharing a sort of wager on which brother would make the first sound. Finally, it was Wallace who broke the silence. He stood up, fitted the heavy package with some difficulty under his arm, lifted his gas lantern, and shuffled toward the door.

"Ah-hanh," he said.

"Ummm," said Tim.

No one else spoke. Wallace opened the door to look out into the night. There was only faint lantern light in the bunkhouse, which was known as the barroom, and some clumping of booted feet down there. Wallace stepped down a step, put his lantern down, then reached back and silently pulled the door shut. We could hear the breath of his lantern diminishing down across the yard. Paul Cyr lifted his head now and studied Tim for a while. Tim had found a piece of newspaper, had put his dark-rimmed glasses on, and was studying the print intently. Paul had sat somber and silent all the time Wallace had spent in the shack. Now he looked up and fixed his brooding brown eyes on Tim, who crouched behind the little lamp that provided all the light for the cabin.

"What say, Tim?" Paul said in a louder tone, "I go to CanaDA next week, run in thirty quart?"

These were prohibition times, when the packing of liquor across the border was illegal and highly profitable. Whiskey that sold for a couple of dollars in Canada brought six from bootleggers on this side of the line, and they retailed it to "sports" for up to fourteen dollars a fifth, although a guide might buy it for ten.

Tim removed his glasses and squinted at Paul across the top of the kerosene lamp that lit the cabin. His long, scarlet face screwed itself up into an intensely sour expression, meant, I suppose, to indicate puzzlement. His eyes, under puffy lids, turned to slits, and his tight mouth grew tighter still as if he were about to spit. Paul's voice rose high, so that he sounded rather like a small boy asking permission to play a rough game.

"I go tomorrow, be back Sunday," he said. "I pack him down over montang. I hide her by old Tague camp up here, never bring her into camp at all."

Tim carefully folded his glasses into a leather case, then cocked his head a little and thrust his lips out thoughtfully, his eyes still squinted. Fingering the short growth of whisker on his chin, he stood up and gave his head another quick twist to suggest the complexity of the problem.

"Well, I don't know," he murmured, with his voice down deep behind his fourth shirt button. "I don't know what to say." There was room in our little tar-paper shack for a man to pace no more than three steps in one direction. Tim walked three steps one way and three steps the other.

"I hide that pack good," said Paul, in a voice more than ever like a little boy's. "I don't bring her *near* that barroom. Them goddamn Dutchmen, she don't even smell that booze!"

Tim accepted this assurance with a nod and a frown. He

seemed to chew on it, swallow it, then bring it back up to roll it in his mouth once more. He squinted at the roof boards. He nodded again.

"Well, I suppose," he intoned, portentous as a judge, "if we can be sure you don't bring the stuff into camp, and if it don't get around that you got it."

"Oh, I don't tell a fuckin soul," Paul breathed piously, getting to his feet. The lamp threw his enormous shadow so that it blackened one half of the shack. "I swear to Jeesa Chri'!"

Tim sucked in his cheeks, nodded five or six times, and let his eyes open again to their natural width. He had walked almost up to Paul and turned away now to go back to his seat behind the table. "I guess there'll be no harm."

Paul grinned suddenly, his snaggled teeth glistening in his black beard. His voice rolled out in its normally boisterous roar.

"And I bring back a quart for you! Hey, you old sunnamabeetch?" With that, Paul swung one mighty leg and landed his right boot squarely in the seat of Tim's heavy trousers, lifting the little man three full inches into the air. Tim came down against the table, rocking the lamp, which I jumped to catch hold of. Tim tried to laugh but managed only a short cough. With his dignity so abruptly scattered to every dirty corner of the shack, he could not even look Paul in the face. He slid into his chair and sat shaking his head, while Paul rattled the windows with his laughter. The scaler and I had gasped in unison when that big rubber-soled boot took hold of Tim's ass and hoisted it high. Now we snickered weakly, not daring to invite the venom of Tim's eye. Paul shambled out the door, which he had to bang twice behind him to make the latch secure. He continued to laugh as he stumbled down across the dark dooryard.

One might have thought, from their having worked in partnership some twenty years, that Tim and Wallace McCormick would have been devoted to each other, as much as brothers ever were. But in truth their relationship seemed brimming with hostility, never openly expressed to each other, but spilling over sometimes in complaints to the rest of us. Tim carried a constant grudge against Wallace's wife, whom he never called by name. Apparently her presence meant to Tim that the take was being cut into thirds, rather than into halves, as by rights it should have been. So Tim took care to check on purchases of food or kerosene or soap or candy to see if any were diverted to the "other camp." And if any were, or if he suspected they had been, he would vow to even matters by ordering an equal quantity of fruit or candy or nuts that "we can chew up over here."

Wallace habitually came to our shack — the office shack — of nights to count over, nickel by nickel, and penny by penny, the money in the petty-cash box. And when Wallace left, Tim would hurry into the little room where the wangan — candy, tobacco, hats, pants, wedges, axe blades, axe handles, and saw blades — was kept and take his own count of all the silver and copper. How he could have known if Wallace had indeed slipped fifty cents into his own pocket or not, I could never figure, for Tim never had the patience Wallace had about checking all the entries in the daybook and all the check stubs, paid outs, and pay orders. Still he dropped those small coins one by one through his fingers, grim as a policeman, perhaps expecting some major pilferage to reveal itself suddenly like a false bottom in a barrel.

Tim, who was every sixteenth of an inch as stingy a man as Wallace, never did order any supply of goodies for us to "chew up." Instead he nursed his bitterness day by day and expressed it in sudden angry gestures or long strings of villainous curses, often so foul as to frighten the simple-hearted Dutchmen who felled the trees and piled the pulpwood. Sometimes they would be so struck with terror when Tim let go with some variant on his favorite "Dirty dying Jesus Christ!" that their eyes would swim in tears, as if they stood helpless on the deck of a ship toward which God's torpedo was already speeding.

Tim would never curse Wallace's wife to her face, for he never spoke to her at all, except to acknowledge a question of hers with a grunt. But when she had gone away, after providing some housewifely aid to Tim, such as remaking his bunk with linen sheets, or gathering up his wool socks to wash them, Tim would fly into such a fit of temper that no one dared stand close to him.

When he found the sheets all neatly fitted to his bed one night, with the dirty gray blankets rolled back and a collar of pure linen edging them, and a snow-fresh pillow, all plump and squared, laid atop, Tim tore the sheets free without regard for how the bedclothes were sent scattering. His face on fire, his lips grim as an executioner's, his breath whistling through his nose, he yanked the clean sheets loose and flung them to the cabin floor, then ripped off the pillowcase and hurled that after them. Only when he had the linen all kicked into a heap by the door did Tim stop to give himself over to cursing. The scaler and I were well out of Tim's reach, I lying on my own bunk above Tim's, and the scaler laid out full length on his own, against the opposite wall. But Tim's bitter obscenities pelted us both, as if they had been flung at us by an angry fist. I cannot begin to recall all the nastiness that Tim dredged up to paint a nightmare image of his sister-in-law. He concentrated on her naked body, detailed all its most repulsive qualities, then consigned them, wholesale and severally, to the deepest pits his vocabulary could encompass. Harold and I dared not

13

laugh, although the image of that gnomelike figure in a cheap army-type hat, his face empurpled with blood and his fists rising and falling in a sort of barbaric ritual, would have wrung a smile from granny's ghost. After a time he settled himself enough to put his bed back together and to make himself ready for the night. But even after he had blown out the lamp and lay snug beneath his greasy blankets, still clothed in his long underwear and heavy wool shirt, the curses kept bubbling out of him, regularly as a snore.

The next day, Mrs. Wallace came by and gathered up all the cast-off linen without a word to anyone and just a slight pursing of the lips to indicate her dismay. Tim by this time was far off in the woods, where, according to the informal arrangement that seemed to exist between the brothers, he spotted out roads for the choppers to follow and generally oversaw the swamping of the roads and the piling of wood where it would be accessible to the teams that would start hauling when the snow grew deep. Harold and I knew, from having watched his plodding figure move through the woods, axe on shoulder, as we stood far off unseen on a knoll, that Tim often took himself a half mile south of our works to where some summer resident had long before built a small springhouse in the woods. Under the skimpy shelter here, Tim would light his pipe, out of the wind, and sit quietly puffing blue smoke for an hour or more.

Wallace meanwhile labored without letup in the dooryard, and when Tim would return from the woods, Wallace seemed to make an extra show of hustling about his own chores — just as if Tim were the schoolmaster and Wallace but a small boy who had been set some tasks to complete for punishment. Wallace was a deft man with an adze, with which he would hew out sleds to be fitted with steel runners. He could shape the steel with forge and anvil and hammer, fix it to the sled, and bolt it tight, with the bolts I had brought him. When the lake was open he had run the motorboat to fetch the freight, had landed and stored it. He ordered the supplies, paid the bills, brought the horses in on a scow, and rationed their feed — as he rationed that of the dollar-a-day boarders, the lumberjacks themselves.

Indeed, there was not a man in the camp who did not work harder than Tim did. But Tim did a special penance that Wallace avoided. Tim took his meals with the rest of us, in the cook camp, dipping his meat out of the common pot, drinking the ink-black tea, and sharing the pans of dark-boiled potatoes and cabbage. The food provided, for a dollar a day, was but barely fit for human fodder. Only two full meals were served in the camp, for the woodsmen carried their lunches with them into the woods, where they built fires to boil their tea or coffee.

The meat served at supper (dinner was the name of the noonday meal) was always some form of beef butts. These were purchased in small wooden boxes that, to look at them, would not have weighed over forty pounds. But the little boxes weighed a hundred pounds apiece; the meat they contained had been so compressed — the juice having been squeezed out for soups before it was marketed — that it had acquired the specific gravity of lead, or something close to it. I know when I first set out to pick up a box, consigned to our camp, from the railroad station platform, my first tug had not even budged it. Several gawky adolescents of my own age and general appearance, who had been watching me, burst out laughing then. This drove me to such a surge of effort that I hoisted the box with one quick heave right to my shoulder and walked off with it, leaving all the laughers (I imagined) undone.

But putting even small chunks of this beef into the stomach required determination of a different caliber entirely, and it was more than I could summon, after I had swallowed the stuff once or twice. The meat was cooked up in a pot full of lard substitute, dug out of wooden pails with a greasy paddle, and set to bubbling on the stove. When it was "done," the whole mess was turned out into pannikins that were set about the long board tables for the men to dip into. If you waited too long to take your share, the grease would have turned back into a gray-white solid that neither Harold nor I could quite manage to consign, unmolten, to our insides.

The boiled potatoes, set out in pannikins like the beef, were always the color of waterlogged snow, for they had remained too long in the pot after being taken off the fire. They were edible, however, as the cabbage was. Some of the Dutchmen even found the stomach to flavor their potatoes with the grease from the meat. It perhaps need not be said that there were always two or three lumberjacks laid out in the barroom, suffering from diarrhea of varying virulence.

The drinks were tea, coffee, evaporated milk, and water fetched from the spring in buckets. The cookee, carrying either the teapot or the coffeepot or the water pitcher, walked up and down the aisles between the tables and filled the pannikins as requested. There were no handles on those "cups," so many of the men drank with their thumbs immersed in their drink. The tea could not be distinguished from the coffee by looking at it or smelling it. Both were black and bitter as tar. But there was always a string wound around the handle of the teapot, so the cookee would know what he was carrying. Most of the men would mix their drinks without regard, taking tea first, then filling the pan with evaporated milk before the tea was quite gone, then taking water on top of the remnants of that mixture, to which sugar had been liberally added at the start. There was little talk at the table, except as the men called out to remind the cookee that a pan was empty, for every man devoted himself to eating, which was almost the only recreation the camp afforded.

Harold and I found ourselves eventually living on cabbage, turnips, potatoes, cookies and tea, with bacon in the morning that had many of the qualities of rock salt, but that was at least more palatable than the beef. The others, even Tim, who had surely known better fare, consumed whatever they found before them, not always without complaint, but always with appetite. When they found some dish difficult to stomach, there was always a jar of molasses to garnish it with and render it sweet enough to swallow.

The boarding arrangement was difficult for me to understand. In all other live-in jobs I had known about, the "found" — that is, the board — was included in the wage. But here the woodsmen seemed to own the status of independent contractors. They received two dollars and forty cents for a cord of spruce or fir, cut four feet long and piled "at the stump" — that is, along the woods road they themselves had to swamp out of the forest. A woods cord in that day measured four feet deep by eight feet long by four feet and four inches high — the four inches being allowed for "loose piling." It was, however, just a means of slicing a tiny bit of extra profit from the woodsman's wage, for the lumber company, when it sold the wood, granted the buyer no extra lagniappe for loose piling. The woodsmen also rented their tools from the subcontractor — the Mc-Cormicks — so in addition to the dollar a day that was subtracted for their board, a fee for "use of tools" was also figured in, the fee varying, depending on the length of time the man had worked. Occasionally, when a man was paid off who seemed less woods-smart than the others, the crafty brothers would credit his tools back at a price somewhat less than they had been charged out. Tim would pick up the returned axe and weigh it carelessly in his hand as if looking for a place to discard it.

"Look at that," he would growl. "Not worth a shit now!" The saw and the sledgehammer (listed as a "swedging hammer") would likewise be cried down, and the whole lot taken back at less than half what they went out for. Then the two or three dollars for "use of tools" would be subtracted as well, while the tools, after a few passes with a whetstone, a file, and a dirty cloth, would be charged at full price to the next man.

It became clear to me that the McCormicks were really running a sort of boarding house and rental service, as well as taking a profit on the wood. They also sold, of course, candy and clothing and tobacco to the workmen out of the wangan, and sometimes apples and playing cards, so there was many an extra penny turned from those stingy wages. The men were required to swamp out their own roads — that is, cut the stumps down so that the snow would cover them and the sleds could ride over them, clean out the brush and pile the tops out of the way, leaving a way clear of brush, stumps, and saplings where a horse-drawn sled could haul a load. The main roads, called two-sled roads because they had to be wide enough for a rack set on two sleds and drawn by two horses, were swamped out by men hired at day rates. The piecework lumberjacks had to stamp their wood too, marking each stick on both ends with a small stamp axe that bit a cold brand into the wood designating the company to which it belonged. The Brown Company, with whom the McCormicks held a contract, used a letter H as their timber mark. Inasmuch as the mark also had to be cut into some timber with an axe, it was desirable to use a mark that could be made with straight axe strokes — an H or an E or an I or a Y. The smaller sticks, with the mark indented at each end, could be identified if they strayed from the drive and lodged on a riverbank or became mixed with wood destined for another mill. The boom logs, cut twenty to thirty feet long, had their brands cut into their bark with an axe — and an able man with an axe could mark an H in three quick strokes. It was also a relatively easy matter for a practiced eye to judge the cubic contents of a log at a glance, a skill in which I prided myself. The lumber mills that shared these waters used them as if they held private title to lakes, rivers, and brooks, and could dam them, jam them full of wood, or turn them loose to suit their own needs and God help those who might thereby be discommoded. In my day, and for long afterward, vacationers, hikers, fishermen, hunters, bird-watchers, and campers were mere nuisances to a lumber company, folk who sometimes had to be chased away from a stream, waved away from a towboat that was drawing a boom full of pulp down a lake, or warned to take their canoes out of water being used for driving pulpwood to the mill.

No one dared question the right of the almighty lumber companies to treat all the waters of the state of Maine as their own preserve. Even at the paper mill, the entire river might be used to wash the waste away, rendering the river for miles below undrinkable, unswimmable, and unbearable. To the people who dwelt in modest homes along the banks, the presence of a noisome stream was simply one more of the prices men paid for being poor.

The McCormicks themselves had been raised poor but they had only scorn for men who had remained that way. The young Nova Scotians they imported to work in the woods were all of German origin, inhabitants of a tiny corner of Lunenburg County, where Germans had settled more than a century earlier, and they all wore names like

Ernst, Aulenbach, Wentzel, and Veinot. In our camp there may have been five Aulenbachs, and none claimed kinship with any of the others. Indeed, when I mistakenly suggested to one yellow-haired young Aulenbach that another Aulenbach, also yellow-haired and also built square as a woodbox, might be his brother, he angrily corrected me. But there was no question that all were damn poor. Their pants were heavily patched and their boots too often worn right through. Nearly every one of them, when he had been in camp two weeks or more, used up a share of his scant earnings in the wangan, buying new pants and "rubbers" to last him through.

A lumberjack "tried on" his pants by closing his fist and shoving his forearm inside the waistband. If the pants fit the forearm tight, then they were the right size. One fellow, laughing as he chose a pair to fit, kept demonstrating his need by lifting one leg and showing how his current pair were completely gone on the inside, from crotch to knee. That any man should exist in such penury aroused open contempt in both the McCormicks. They were soft-spoken as could be in dealing directly with the men, but out of their hearing, they spoke in disgust of the way the Dutchmen ate, slept, washed, and relieved themselves.

Their manner of performing most of these simple tasks was largely dictated by the facilities the McCormicks provided them. But there was no doubt they had all been brought up in homes where the only plumbing was installed at the well, where all natural needs were attended to outdoors, and where laundering was a luxury. Here in the lumber camp, they all slept, half clothed (they'd have frozen otherwise), under a lumpy fat "comforter" made of horse blankets sewed together and stuffed with cotton waste, in a straw-filled bunk, two men together in a bunk about three feet wide. The straw, of course, frequently crawled with live creatures, which moved freely from one body to another all up and down the double row of bunks. The barroom, or sleeping quarters, and the cookshack were all of a piece, a long low building made of used boards covered with tar paper. A door and a high sill, over which sleepy lumberjacks frequently stumbled, connected the two large rooms.

Some twenty yards above the bunkhouse, out of sight behind a clump of alders, a pit had been dug and a lean-to shelter rigged over it, open at the sides, with a long pole stretched across the pit at just the right height for a man, standing on a narrow strip of planking on the edge of the pit, to lean back and hook his arms over. There was room for three or four men at a time. This was the spot provided for defecation. It stunk to the skies. Tattered bits of newspaper or advertising circulars, used as toilet paper, formed a sort of carpet on the planking or blew out to adorn the brush and snowbanks all around. There was a bucket of loose earth at hand to be tossed into the pit atop each new deposit, but few men out in the zero cold of an evening paused to perform this office. "Ain't as if they was *human*," Wallace would snarl occasionally when he chanced near the place and shrugged off its foul condition.

The kitchen staff consisted of cook and cookee who slept in two bunks behind the stove. The cookee's assignment was to cut stovewood and keep the woodbox filled, to start fires, to wash pots and pans, to set the table, serve the food, keep the floor clean, and the sugar bowls filled, and put oil in all the lamps. It was also his duty to waken the whole camp in the morning, at half past five, and to call them all to meals. This he did with the traditional wailing cry that cookees had sung out over pitch-dark dooryards since men first gathered in gangs to fell trees and bring them to market.

"Turn a-a-a-a-a-out!" he would call, first into the sleeping barroom, where the fire had died to a few blinking coals, and then out across the yard to wake up the office crew. The sound of this strange cry, disembodied as the call of a distant loon, and echoing like a loon's across the dark empty lake and against the wooded mountain, is one that haunts me still. It rose and fell like the cry of a sailor at sea, seeming, despite its throatiness, as gentle as music. "Turn a-a-a-a-a-a-out!" I would hear it first, muffled as in a dream, inside the walls of the barroom, and then, some thirty seconds later, instant and clear, just across the yard from our door. It would be black dark. The sound of men thumping about in the barroom to pull their boots on would follow instantly. Harold, who could reach out to the table without leaving his bed, would, after three or four attempts, set a match to spluttering into flame, and would light the small lamp, suddenly isolating our little cobwebbed world from the vast cold night. Then, after a few quiet curses, he would hop out in his socks to open the stove door, set the drafts, shove in new wood, and get the fire to pulsing.

We had wash water in a kettle on the stove and a small basin to wash in but often we would first take a turn trudging, with boots unlaced, out to the small, snugly boarded shithouse that was our very own, with a hole in the seat and a board cover to go over it. It would sometimes be just cracking dawn, so that I could make out the men who stood outside the barroom door, all facing the east in a semicircle, like worshippers, and all urinating solemnly into the snow.

Tim McCormick, whose scorn for his fellowman was more caustic than his brother's, or at least expressed more openly and more often, found frequent excuse to share his conviction that all the young Dutchmen were engaged, night after night, in "whetting." That was Tim's word for masturbation, an activity that Tim, like practically every other two-fisted male alive in that day, devoutly believed led to cowardice, distaste for outdoor labor, muscular flabbiness, loss of appetite, and eventual homosexuality. There seems hardly any doubt that these young men — but one or two of whom had gone more than a short stride past thirty, all of whom owned less worldly wisdom than a modern eight-year-old, who were largely married and all vigorous and emotionally stable, and every last one of whom lacked the brass to offer a half-day's pay to one of the scrawny whores who operated semiprofessionally in Oquossoc — certainly required and undoubtedly found some occasional sexual outlet. There was no homosexuality in the camp. If there had been, it could not have remained secret even half a day. And the brothers McCormick would have rooted it out as they would have driven out a thief — by instant violence and consignment to a cold hike across the ice.

So Tim took regular satisfaction in listing this failing among the myriad that he and Wallace regularly ascribed to the bulk of their hired hands. As for the hands, they accepted scoldings from Tim sullenly but without any overt response, other than a few mutterings or an invitation to "go to hell already," offered when Tim was well out of earshot. Wallace, who had not yet gone gray, as Tim had, somehow maintained a reputation as a ready man with his fists, although none I ever met had once seen him in combat and a few had vowed they would welcome a chance to try him. The Dutchmen, however, lived in mortal terror of Wallace's anger and would no more have offered to have it out with him than they would have challenged Paul Cyr to a set-to with pulp hooks. But their resentment built inside them every day, with every meal they took at his stingy table, with every night they spent in his louse-ridden bunks. And when it burst out it did so in a fashion altogether suited to their childish hearts.

The first Sunday after the ice had been found solid enough

23

to hold a man, or a whole troop of men, the thermometer having fallen well below zero and stayed there nights on end, all the Dutchmen in camp who were not laid low with diarrhea took off across the lake for Oquossoc. In that tiny grouping of small frame houses, shacks, cabins, and workshops, only the Catholic church and a two-story lunchroom-boardinghouse known locally as the Dead Rat offered public accommodation. The barbershop did not function in the winter, nor did the Protestant Community Church. Mackenzie's General Store would tolerate only a certain amount of loafing about under guise of waiting for the mail. And the railroad station opened for business only long enough to deal with the one train that came in soon after dawn and left at suppertime. The Dutchmen confined their carousing, therefore, to a half dozen bottles of home brew shared in the back room of the Dead Rat, to a bag or two of candy bought at Mackenzie's, and to some high-pitched clowning performed mostly to engage the attention of the teenaged daughter of the man who operated the general store. They all sent postcards home and one or two bought mittens of a better sort than our own wangan offered. They came back in the dark, marching in a scattered squadron over the wide ice, and forgetting, as simple folk often do, that voices carry over ice or water three times as far as they might in the woods. Every word they shouted to each other rang clear and loud to all of us in camp as if it had been spoken into a telephone. Tim and Harold and I, all attending our own affairs in our tiny shack, laid down our books and papers and paid single heed to the voices far out on the lake.

"The old bastard's light's out," one of the Dutchmen called. "He's giving it to the old lady already. I bet he's giving it to her mad!"

The next man was bolder.

24

"He's sticking it into her!" he yelled gleefully. "He's giving her that old pickpole!"

After that, they all contributed comments and urgings in exactly the same vein. Perhaps only a man raised in that innocent day can imagine the deep sexual thrill these boys derived from yelling out their fantasies so freely.

"Come on, Wallace! Give her that old cant-dog handle already. Stick it in a mile already!" "I bet she got a box like a canoe already!" Almost strangling with laughter, the boys tried to outdo one another in farfetched obscenities. Then they all fell silent together and we heard no other sound until nearly half an hour later, when they banged into the barroom and whooped there awhile, working off the last vaporings of the home brew.

While the cries had come over the ice to our ears, Harold and I had been too aghast to look straight at Tim, who had sat erect to listen. When the obscenities had at last ceased to echo on the mountains, Tim got up to spit into the stove.

"I don't know why they don't take a suck at each other," he snarled, with so much venom that Harold and I could not even pretend to laugh at him.

Next morning, before breakfast was over, Wallace appeared in our dooryard, carrying his lantern, for the sun had barely whitened the eastern sky. This was the first time he had ever appeared on our side of the brook before daylight. We watched him go by the window and heard him enter the barroom through the outer door. All he did in there was take a seat on the bench that ran the whole length of the tier of bunks, and wait silently as the men came back from breakfast to make ready for the trip up the mountain into the pulp works. Always the first to go was Paul Cyr, who worked alone, with a bucksaw, and saw no need to dally. This morning Paul stood for an extra few seconds at Wallace's side, waiting to learn what brought him into the barroom so early, but Wallace merely offered a wry joke of some sort about having got tired of resting and Paul took off, with his lunchpail and his double-bitted axe and his bucksaw. Once he was safely on the road up the mountain, and the Dutch boys were all back from breakfast, a few of them warily watching Wallace, some talking more loudly than normal to indicate that they hardly noticed his presence, Wallace got to his feet. All the Dutch boys, even those who were pretending they had not seen Wallace at all, fell immediately silent.

"Now," said Wallace, in his mild, slightly nasal voice, "if the feller that had all the funny remarks to make about Mrs. McCormick last night will step out here in the dooryard, I'll *intertain* him!"

Nobody spoke.

"One of you sons of bitches had a lot to say out on the lake last night. Now if they's an ounce of manhood in you, you'll step up here and face me." Wallace's eye fell upon one of the Dutchmen, who had looked up from lacing his rubbers to watch Wallace. The young man's face had gone white as milk. When Wallace's glance touched him, he dropped his eyes at once. But Wallace aimed his talk directly at this one. "You was out there last night, Freddie," Wallace declared. "I know goddamn well you was. Maybe you'd like to let us hear what you had to say."

Freddie, so scared he seemed to have shrunk inside his shirt, looked back at Wallace without quite raising his head.

"It wasn't me," he whispered. Then Wallace stared at one and another of the two dozen young men, most of whom quickly turned their eyes away from his. There were several feeble high-pitched replies to the question Wallace seemed to be asking. "I never said anything." "I was here in this camp. Cookee seen me."

Wallace picked up his lantern, which made more noise

now than all of the men together, and addressed a comment to the entire room. He squinted his eyes tight.

"Goddamn funny where all those big *heroes* got to so sudden. If I ever see a collection of *shit* . . ."

His mouth tight, then, he turned and went out the door, leaving dead silence behind. After the sound of his lantern had faded, there were some giggles in the room. But most of the men seemed too ashamed to speak. It was not until they were outdoors, trudging together in the dusk, that they muttered feebly to each other and tried to laugh.

Paul Cyr did not learn of this episode until he had come back at dark, and Harold told him, Harold and I having heard it all from our post just inside the connecting door between cookshack and barroom.

"Sunnamabeetch!" Paul gasped. "Goddamn! Hey, if I be there, by Jees' I confess! You betcha my ass! I confess!"

This was as close as Wallace ever came, in my knowledge, to fighting any man with his fists. And none of us rendered him much credit, for we knew how faint the chance had been that one of these simple-hearted young men would have stood up to bleed for his right to use dirty language in front of ladies. It was just as if Wallace had strode into a schoolyard at recess and dared the entire eighth grade to send him out a champion to fight him to a finish.

But Wallace and Tim both now had one new charge to lay against the Dutchmen. (No one ever called them Canadians or Nova Scotians, least of all the McCormicks, who had themselves been raised in Lunenburg County.) And all agreed that there were few spots in the hemisphere more godforsaken than the village these boys grew up in. The Dutchmen, however, owned to no shame for being poor. Although they wrote few letters and received little mail, their homes and their families were continually in their talk.

And at times their homesickness would burst out like a sudden fever.

Young Bill Veinot one morning had the bad luck to trip and fall over his sledgehammer as he turned to flee from a tall spruce that had taken a wild twist when it was felled. Bill was caught under one long limb and flung to the ground beneath it, where he lay screaming, with his left leg twisted under him. It took three boys chopping madly to set him free. Then when Tim came up and they tried to get Bill on his feet they could see his leg was broken. They made splints out of a spruce bough, and lashed them to his leg with their belts. As they carried him over the rough and slippery ground to the camp, he fainted.

Fortunately the ice was thick enough then to hold a horse and sled, so Bill was given a fairly comfortable ride over to the other shore, where Wallace stowed him and two of his mates into his open Chrysler and drove him to the doctor's.

Bill was bedded down in a boardinghouse in Rangeley, where the Dutch boys took turns visiting him on Sundays. After two weeks, with the McCormicks fretting the doctor over the cost of keeping the young man in idleness (they had hastened to "pay his bills" lest he seek some more extensive redress), the doctor agreed that Bill might take the train home — or several trains, for he would need to shift from the narrow gauge to the standard train that would bring him to Megantic, where he would change again for home. Obviously he would need someone to lift him from one car to another, so two of the Dutchmen volunteered to go back with him.

The word spread through the camp in less than a breath. Bill and Ned and Fran were going home! It had been about four months now since any one of them had seen his own fireside, and the recollection of it must have appeared like a vision before all their eyes. Because within an hour four more lads had come up to the office camp to ask that they be promptly scaled up and paid to date. They were going home with the others!

Wallace was dismayed at this wholesale desertion. The loss of seven boarders might mean inability to meet his contract terms before the snow grew too deep for the choppers. He sat in the little wangan room, fountain pen in hand, staring at the blank boards in front of him as he tried to talk the men out of leaving. He would have to charge them "extry" for carrying them out. He couldn't give them full price for wood they hadn't stamped. They was some roads wasn't fully swamped. . . . Now if they could hold off until the end of the month . . . There was no room for all the men to stand beside the shelf Wallace used for a desk, so some crowded tight in the doorway. None of them made reply. Now supposing, said Wallace, he took them all out to bid good-bye to Bill Veinot, and gave each one, instead of a company order for all he had cut, a small check he could send home . . .

Peter Ernst, the tallest of the Dutchmen, a man with cheeks so rosy-red and hair so blond that he seemed as if he had come out of a little boy's picture book, stood with his mouth drawn down and his eyes focused far away.

"That isn't what I said I wanted," he intoned. "When I came in here." The singsong, childish manner in which these words were uttered caused Wallace's lip to curl involuntarily. He caught his own angry words before they passed his teeth, however, and just worked his mouth around a little. Finally he shook his head and looked out, over the shoulders of the men crowded in the door, to catch Harold's eyes.

"You got them scales?"

Harold passed the papers over to Wallace and I handed in my own reckoning of how much wood had been cut by each crew.

"This all stamped?"

Harold looked at me. We both knew that there were some six piles unstamped, but I had determined to stamp them myself come Sunday.

"They're stamped," I murmured. Tim had been sitting on his bunk, listening to every word. He stood up now and shuffled over to be near the door.

"Seems to me," he said, in a sort of pretend-innocent tone, "they was *some* piles never been stamped, out where these fellers was working."

"I made a deal with them," I lied quickly. "I'm going to stamp them Sunday."

Wallace scowled at me. But my time did not belong to him on Sundays and he had no ground at all for complaint. He took the papers and began laboriously to figure out the dollars and cents value of the cords. When he had the first paper done, he offered a look at it to Peter Ernst.

"I took off for the transportation," he said softly. "And I had to charge you for that axe handle."

Peter flushed.

"Axe handle? I never . . ."

"All I know's it got broke somehow," said Wallace, in the same innocent tone that Tim had used. Peter waved one hand.

"Make her out!" he growled.

I leaned forward to get Wallace's attention.

"Peter and Fran cut that boom log," I said.

Wallace's face turned bright red and he bared his teeth in sudden rage, like a dog, glaring into my eyes.

"That's right!" Peter shouted. He tapped one finger on the page. "That should be down there already!"

30

Wallace had to stare silently at the open order book for several seconds before he could swallow all the words that boiled up in him. He dealt me one final venomous glance before he could control his voice enough to ask me:

"How many cubes?"

"Thirty-six," I said. Wallace picked up a long brown pencil, wet it in his mouth, and painstakingly turned caliper cubes into cords and then into money. He let Peter see the final figure, accepted his nod, and carefully wrote out the company order — one stub, another stub, and then the order itself, drawn on the Brown Company and cashable at Mackenzie's store at a two percent discount. Wallace detached the stiff blue paper with tender care and passed it over to Peter, who read it with equal care, then folded it in two and packed it into the small snap purse he carried. One after another then each man stepped up and received his money. None looked happy at the figures. Each shook his head gloomily, and exchanged brief grimaces with the others. When they trooped out of the cabin, they began to growl to each other, but their words were not loud enough to be heard.

Wallace and Tim took counsel together, muttering so that neither Harold nor I could make out much of what was said. When Wallace came out, he stopped to glower at me.

"In this business," he said, in that light-footed voice that seemed wrung pure of all rancor, "a man don't pay out more than he's asked for."

"Well," said I, feeling my face grow red, "I thought it was only right . . ."

"He *thought!*" Tim exclaimed.

"Well, this is *business*," Wallace said to Tim, still using his tiny, fake-innocent tone. "He wouldn't *understand* that."

I said no more but just blushed hotter still. Wallace and Tim went out together to make sure, I suppose, that the men did not carry off any treasures with them. Harold watched them out the window.

"What a fine pair of pricks!" he said.

As we stood there, we spied Peter Ernst hurrying back to our cabin and we both instantly began to look around to see what it was Peter might have forgotten. But he had left nothing behind. He burst into the cabin, half out of breath, and stood before me.

"We want to give you this!" he blurted out, digging into the pocket of his coat. "You done me good! You done us a favor!" In his open hand he offered me a large compass, the size of an old-fashioned pocket watch. The case had once been silver-plated, but most of the plate had been worn off from much fingering, showing the yellow brass. I drew back and tried not to take it, but Peter grabbed my arm and tried to shove it into my pocket.

"Take it already!" he insisted. "It works good. It's got a jewel bearing."

"Well, all right," I said, taking the compass in my hand and snapping it open to see the intricately decorated face. "I can use this in the woods all right."

"Leastways you won't get lost already!" Peter shouted. Then he plunged out the door and ran to join the rest of the party, who were already out on the ice.

❧ 2 ❧

WHAT may have prompted the McCormicks to fire the
cook was the increasing number of woodsmen who
came down each day with diarrhea. Or it may be that Tim
McCormick, despite his apparent indifference to what he
put into his stomach, had himself begun to wonder what
the steady diet of beef remnants and lard substitute was
doing to his insides. Neither Tim nor Wallace seemed to
own even a normal respect for his own vital organs. Wallace,
indeed, when he had either comfort or distress to report in
the region of his stomach, always referred to it as his dung
bag.

But Tim returned one day to the camp in midafternoon
to confess to having suffered a "falling-out spell" in the
woods. Sudden dizziness had forced him to grab hold of a
popple trunk and slide gradually down on his pants in the
thin snow. He had found himself sweating so then that he
took his hat off and grew frightened to discover that his
very perspiration reeked of tobacco, as if it was inching
through a dirty pipe. When he had the strength, he took the
pipe out of his pocket, he said, and threw it the hell away

33

into the brush. Then, his equilibrium gradually returning, he made his way by slow stages back to the cabin, where he lay flat on his bunk with all his clothes on, too woozy now to even try to revive the fire.

I had been selected that day to help Wallace erect a sort of ell or "baker" on the small blacksmith shop, and had spent most of my time laying the roof boards and tacking down rolls of tar-paper roofing. So I had found Tim abed and heard his story first, and had got a fire going to keep him warm. When I brought word to Wallace he dropped his handsaw right on the snow, an act of incredible slackness for him, and hurried over to the cabin. I followed and found him perched on the little bench that ran alongside Tim's bunk, not looking into Tim's face but offering to the empty cabin the most doleful expression I ever saw him wear. He straightened his face somewhat to look up at me, managing a small frown, as if I might somehow have been implicated in what had befallen his brother. But the frown faded instantly and he turned to Tim, who sat up when I came in and got his feet out over the edge of the bunk, as if he would rise up at once. He merely sat there, however, braced on his two hands.

"That Jesusly tobacco," Tim said throatily. "What in Christ's name ever gets a man to sucking that shit into his guts is more than I know."

Wallace pursed his lips tightly and moved his head to one side.

"Now, I don't know," he said, in a voice so light and faint that it made hardly more sound than a whisper. "Not the little you smoke . . . Just the pipe . . ."

"It gets into your system," said Tim. "It just gathers in there and soaks your whole system. It come out of me like pitch out of a pine log."

"Still," said Wallace, shaking his head as he spoke, "seems

like it would take something to *start* that. . . . You wasn't smoking. . . . Now if it was something you took into your *stomach* that morning . . ."

"Well, Christ," said Tim. "I et the same as I always do. What I really need is a good tonic, to clean out my system."

Wallace frowned deeply at this and spoke with more vigor.

"Now I ain't so sure," he said quickly. "What is it they tell us about that?"

Tim got up and rubbed himself under the arms and on the back of his neck and said he felt right as rain. Only he guessed he would pass up dinner this noon.

It was just two days afterward that Wallace fired the cook — or rather sent out for a new cook, for it would have been folly to let your cook go without knowing you had a new one at hand. The new cook, a small gray man from Nova Scotia, about the same size and disposition as the Mc-Cormicks themselves, spent the first night sleeping in the upstairs bunk over the scaler's. Tim, who sat on the edge of his bunk swigging Father John's Medicine out of a dark bottle, explained to Harold and me that now we would get a good piece of meat once in a while and not the same shit every day. The cook, daintily sorting through the contents of the small imitation-leather grip he had carried in, looked up with a half smile and said something self-depreciating, but he sounded pleased all the same. Harold and I, in one quick glance, managed to share our skepticism. It would take more than the magic of a Nova Scotia woods cook, we knew, to turn those frizzled beef butts into delicacies. But we had hopes all the same that there might be more tooth-some doughnuts for breakfast than the grease-sodden bits of sweetened rubble the current cook provided.

The old cook was fired the next day after breakfast, amid angry shouts and threats of litigation that had all of us, throughout the camp, frozen to attention. One major complication of getting rid of the cook lay in the fact that we would be short of a cookee too, for the current cookee was the cook's own son, a big, lumbering sort of boy with a way of dodging a greeting by twisting his head to one side and looking off at the ground. The cookee had nothing to say back to Wallace when Wallace quietly told both of them, once breakfast was all done and cleared away, that they were getting through. But he stood there and glowered angrily to add weight to his father's demands that they be both given a month's wages in place of notice.

"By Jesus Christ!" the cook yelled, when Wallace was walking away from them across the yard. "I'll take you to court on this. I knowed you two was tighter than a crab's ass! But when you set out to beat me, by Jesus, you picked on the wrong rooster! I know the *law*, by Christ!"

Wallace made no reply. But when he had come back to the office camp to take counsel with Tim and the new cook, he mimicked in a startlingly accurate way the cook's last cry:

"I know the *law!* I know the *law!*" Then Wallace added in his normal tone: "There's no such a goddamn *law!*"

"Not a goddamn bit of it," Tim assured him. "Maybe kind of a half-assed *custom*, in some lines. No man ever heard of any such a goddamn thing in the woods."

All the same, being litigious men themselves, the McCormicks were not entirely easy in their minds that they were safe from a lawsuit. I was sure Wallace would write a note before long to the Rumford lawyer who from time to time sent them small bits of advice about their rights against a man who had pastured their horses or someone who had eaten up a portion of the potatoes they had stored in his cellar. The lawyer habitually charged them two dollars for such random bits of wisdom. It seemed likely that they would deem it worth two dollars to live free from the fear that the departing cook had any recourse.

Meanwhile there was the obligation to transport the cook to the train station whence he had been fetched. This obligation was one so long honored in the woods that Wallace never questioned the need to conform to it. He returned to the cookshack, where the echoes of the shouting had long since petered out, and solemnly told the two sulking men that he would be hitching a sled in ten minutes or such a matter to haul them out with all their gear. The cook and his son by this time had got into their city suits and their ancient dress-up felt hats.

The cook's son, who was still addressed as "Cookee" by everyone, had put on sixteen-inch lace boots such as artillerymen wore. But the cook had dressed his good shoes in four-buckle city-style overshoes, so he looked as if he had just stepped out of a trolley car into the snow. The cookee, with his hat cocked over one eye and his high boots shining, cut a surprisingly impressive figure. He really was a well-built lad, with muscular shoulders. Except for his mouth, which was fat and formless as a small boy's, and always seemed moist, he might have posed as the very model of a rugged outdoorsman.

Neither the cook nor the cookee had any good-byes to say, for both seemed aware that there was general rejoicing at the prospect of an end to the frizzled-beef regime. They each owned a bulging paper suitcase and the cook carried, in addition, a small grocery carton, also bursting full and tied with a piece of halter rope. They made their way down to the dock and stood gazing out over the glinting snow, as if a magic boat might push its way through the ice to carry them off to a land where their skills would be made much of.

What came instead was a one-horse pung, sent down

from the storehouse, carrying Jim Kidder, a bright-faced young man with the clearest eyes and the heartiest laugh I had ever known. He was a hulking man, yet quick-footed as a spider, well over six feet tall, exceptionally long in the arm. His hands, cased in fingerless choppers' mittens, seemed each one the size of a small shoulder of pork. His job was to check the scale on our job, to make sure the required number of boom logs had been felled and limbed out, ready to be twitched down to the ice, and to confirm that no usable wood over six inches at the stump had been left standing. His happy arrival made it unnecessary for Wallace to hitch up a sled and ride in contentious misery the long five miles to Oquossoc. The pung was headed for the train station too and, instead of taking the winter tote road down over the corduroy bridge, had swung out of its way just to save Jim Kidder the hike.

Only a few seconds' negotiation was needed to persuade the driver to take our ex-cook and his son aboard. They tossed their duffle into the wagon, the cook shared the blanketed seat with the driver, and the son climbed in to nestle down among the luggage. Jim, who had been busy shaking Harold's hand and expressing his joy at making my acquaintance, did not take full note of the departing pair until they were already moving off, harness bells jangling and driver urging his heavy-hoofed steed to get up in that collar. Then Jim Kidder looked after them, as if he had forgotten to say good-bye. He turned to us, with a wide grin.

"I *know* that young joker," he declared, in a voice loud enough to carry over the lake. "Ain't he the one I carried up here in the boat last fall? I know goddamn well he is! Sure he is!"

"He came in with you," said Harold. "The same trip you brought the schoolteacher."

"By Jesus, yes!" Jim shouted. "He was some kind of a

detective, he said. Come up here to take some feller you had working here. A counterfeiter or some such goddamn thing. Christ all Jesus! He had that little girl atwittering!"

"Some detective he was," said Harold. "All the detecting he done was in the dishpan. He was cookee here for his old man. Peters, the cook."

Harold stared after the diminishing pung with a half-admiring smile.

"Why the young pisshead! I believed him. He showed me some kind of a half-assed badge and a pistol he carried. I never did see him again. You know I thought it was goddamn funny I never heard any more. You suppose he come to pinch his old man?"

"His old man ran his ass around," said Harold. "Why he was just as scared of him as if the old man had horns."

It was true that, despite the fact that the boy stood a head taller than his father and must have outweighed him by a good forty pounds, he still accepted his father's abuse in shamed silence, sometimes looking as if he might burst into tears as his father detailed his failings for the whole camp to hear. Yet he seldom spoke to anyone other than his father and in the evenings the two would often sit and mutter and laugh together like cronies in a tavern.

"Somebody told me he pissed the bed a lot," said Harold, and when we both laughed, he hurried to add, "I don't know. It's just what somebody said. I noticed his father took the top bunk and let the boy sleep downstairs."

Jim Kidder set his head back and gave out a roaring laugh that brought echoes from all the hills.

"I sure as hell wouldn't want to sleep under his bunk myself!" Jim shouted. "Not without they caulked all the seams!"

The racket Jim made drew Tim McCormick to the cabin doorway, where he peered out somewhat fearfully at first, like a woodchuck come to the entrance of his burrow. Tim was hatless, in shirtsleeves and suspenders, and held his reading glasses in his hand, but when he recognized Jim he came down the short steps into the cold, offering a sudden grin that seemed to split his whole face. There was no mirth in the expression at all, for it was like the grimace of an ape, a mere wide showing of all Tim's little brown teeth. His small eyes became slits. The effect was anything but pleasant, yet there is no doubt that Tim meant to display the utmost goodwill, for Jim Kidder was the embodiment in our neck of the woods of the great Brown Company, from whom all blessings flowed.

It was the Brown Company that let the contract, that paid for all the usable wood that was landed on the lake, and that advanced the cash to cover the "company orders" with which the woodsmen were paid. They also took care to see that their rules were obeyed as to cutting, piling, scaling, and stamping of wood, as to swamping roads, piling brush, and cutting clear all the spruce and fir in the area over a certain size. Naturally, on this account, Jim Kidder was a potential enemy to the McCormicks, who found the whole world full of enemies anyway. But it was not their nature to seek trouble with any who might do them in. Indeed, they were both such masters of the soft answer and the innocent glance that only those who had known them as bosses or as business adversaries understood their true natures. Jim Kidder was one who had met both Wallace and Tim on several footings, as he told us in the woods that day. He had once, when guiding a bateau along the lakeshore to rescue castaway lengths of pulpwood that had escaped from the boom, found occasion to offer to beat the ass off Wallace, who was given to talking himself up as a man ready to "intertain" with his fists any who dared cross him. This time, Wallace had mistaken Jim Kidder for one of the thoroughly subdued Nova Scotians who came down here each winter to grub for scanty wages. And when Jim did not respond to the hail and bring the bateau in to ferry Wallace across to the storehouse, Wallace had suggested that Jim needed to go back to Nova Scotia and get his ears cleaned out. Or perhaps Wallace might clean them out himself.

Jim Kidder's temper, as I soon learned, lay balanced on the very narrowest edge of his nature. At Wallace's challenge, he changed course with a surge that very nearly dumped the two men, with pickpoles, into the lake. Jim climbed out before the bateau had grounded and waded up to where Wallace stood among the rocks.

"You hairy-assed herring choker," Jim greeted Wallace (or at least this is the descriptive Jim remembered offering), "I'll break your ass in seven places."

Wallace, before Jim even came within reach, had started to move quickly backward up the bank, both hands extended peaceably before him. "Now wait! Wait!" he cried. "I don't want to fight you! I took you for one of my boys! I couldn't make you out too clearly!"

"Well, you make me out next time," Jim urged him. "Or your old lady'll come pick up what's left of you on a pickpole. Now you make your own fucking way back to the storehouse. And stay clear of me!"

With that, Jim related, he climbed back into his bateau and went on about his business of gleaning lost pulpwood, which, while worth only two dollars and forty cents a cord to the men who cut it and stacked it at the stump, was still too precious to the company to let it bake into dry-ki. Wallace stood marooned still upon the rocks, and would have been there yet, I suppose, had not some better-disposed boatman come by.

But all this was part of what Jim Kidder told us after we had left the camp yard and moved up into the works. First he had to stop and deal with Tim, who came down and offered a hand for Jim to take brief hold of.

"Don't be too tough on us," Tim said, laughing as he spoke.

Jim did not smile in response.

"Depends on what you been up to," he said flatly. He glanced at me and winked. Tim accepted this as a joke and laughed again, holding his place before Jim and keeping his mouth open as if to allow more laughter to escape at the slightest signal. But Jim moved around Tim and continued on up the wide trail, where the snow in many spots had been worn down to the rock ledge by the plodding feet of the woodsmen going back and forth. "Better bring your stamp axe," he told me, and I dodged away to fetch it from the cabin.

This being Sunday there was no sound of chopping, nor any lazy whine of crosscuts being pulled to and fro. Red squirrels scolded us for walking into their woods. Chipmunks, which were as common then as sparrows in a city street, fled from almost under our feet and vanished into holes too tiny for us to see. The cutting began well up on the hill, in an area full of blowdowns, for the rocky ledges allowed only shallow rooting and stiff winter winds had torn many a high spruce free of the rocks and thrown it splintering down through its crowded neighbors. The brush piled by the woodsmen was higher than our heads in spots, still not rotted down enough to offer a vista, and covered thinly with new snow. There was mud in all the new-swamped roads. I knew the choppers were supposed to be despoiling this virgin forest by dropping all the tall softwood and leaving a sort of desert of hardwood saplings and brown evergreen sprills to broil in the summer sun. But to me that day the prospect was as beautiful as the vision of a magic woodland. The sky was bright and there was no sharp wind to muffle from. As we topped the first short rise and looked down into the wide ravine, the woods at last lay open before us and we saw ranks upon ranks of pulpwood piled among the small trees with tufts of golden tamarack, tiny firs, and skinny spruces still persisting among the barren birch and popple. Cut spruce and fir are pale yellow at the ends, about the color of the new-risen moon. There was no sense of ravishment or destruction at all, despite the many stumps, freshly healed by the snow, and the sprinklings of saw cuttings, called "cabbage" by the woodsmen, although it looked more like some patent breakfast food. The random piles of wood seemed both orderly and disordered, some running two or three cords on the face, and some, piled narrowly between the small trees, measuring no more than four feet across. The impression was that of a store of natural treasure tucked here and there by a giant hand, with the snow laid thinly over each pile like a blessing.

We three stopped and looked down at the wealth of wood and said nothing for a moment. There was time enough to do all Jim had to do and leave half the day over, so we meant to enjoy the sun and the sweet pungence of newly cut wood and oozing bark. Jim took some time then to tell us of his dealings with the McCormicks.

"Old Wallace has been near the top of my shit-heel list ever since I first knew him," said Jim. "But I put Tim right at the head of it. He's about as miserable a prick as I ever seen in these parts. That right, Harold? You rate him a fourteen-carat pisshead?"

"I put him right up there," said Harold, who was, to be sure, a far more tractable type than Kidder, not given to finding open fault with his superiors.

"Well, I give you an example," said Jim. "The prick bragged to me one day about how he beat a whore out of her two dollars down in Boston. What a Christly thing to *brag* about! Jesus, if I was so goddamn miserable as to pull a shitty stunt like that I'd keep it to myself. Right, Bob?" Jim looked at me open-eyed as if he truly yearned for confirmation and I hastened to mutter agreement. "Why sure," Kidder went on. "But that son of a whore was proud of himself. Near laughed his fucking gut out. Told how he made out he had the cramps and had to run for the shithouse. Then he grabbed his coat and ducked out the fucking door. Old whore lady she opened the window and threw a pisspot at him and near brained the bastard. Wish the hell she had. Can you imagine doing a girl out of her pay that way? If I knew that shithead he made her earn it twice over. Can a man get any lower than that?" He addressed this final question to me in his usual manner, seeming to beg agreement.

"Not much," said I.

Kidder went on about the brothers, whom he had coped with over many seasons, when they cut long lumber, before going to pulp, and when they ran the little steamboat that towed the great log booms down to Upper Dam, where the wood was sluiced through into the river. He was a man who loved to tell stories, and as he went on about the McCormicks his tales encompassed half a hundred strange folk who had peopled these woods in the past. He had a way of assuming that you knew almost any man he might mention, and he would refer to them as "Old Larry" or "Old Wentzel" or "Old Bisbee." Or if he had some doubt of your ever having crossed paths with someone he spoke of, he might offer an obscene characterization as if it described the man well enough so you might know him.

"You know Doc Hanson?" he would ask suddenly. "The pisshead?" And I would have to confess that the description did not help my recollection at all.

We moved among the pulp piles, noting the culls — the portion of a stick where Harold might have marked off a rotted part to be subtracted from the total measure, or the sticks marked with a black crayoned X to indicate they were "pas bonne" altogether. One of the young Frenchmen who had been imported to replace the departed Dutchmen habitually cut popple and piled it in with the spruce and fir where its chalky face stood out from its mates. These Harold would studiously x out and carefully reckon their diameter so their cubic contents would not be credited. Jim noted this with approval.

Jim Kidder carried Harold's rule for a time, scaling a pile here and there and asking Harold to check his own measure. And he would note sticks that had been skipped by the stamp axe and ask me to stamp them. He found a few high stumps and kicked them lightly.

"You got to teach your Jesusly Dutchmen to bend over a little," he told us both. "I didn't think anyone but a Frenchman would cut a tree chest-high."

We moved now into the small yard where Paul Cyr had been working alone, in a hollow, with brush piled high against the wind, for Paul went out into the woods regardless of wind, snow, or rain, built himself a fire there, and cut wood as long as he could see. Jim laughed to see the name "Paul Cyr" crayoned boldly on the piles.

"Old Paul," he said. "He makes that bucksaw smoke. He's like one of those old-time choppers you used to hear about. They used to tell stories about old Toothaker, said, by Jesus, he'd swing that old poleax till she glowed redder'n a stove lid." Jim laughed his rolling laugh. "Said he used to carry that old axe down to the stream to dunk her and cool her off, and, by Jesus, that water would get to bubbling so they could toss the hogs in to scald off the bristles!" Jim laughed again. "That the way you heard it?" he demanded of Harold. Harold shook his head, grinning. "Well, they used to tell some awful stories, back along," said Jim.

We moved next into some fresh choppings, where the snow was soiled with mud, torn bark, and chips, and where huge blue rocks rose up higher than two men. This was the spot Tim had awarded to another bucksaw man, a young Frenchman built like a wrestler, over six feet tall, with long arms and a head that he kept shaven clean. He was a Rumford man who spoke English with no accent at all and only a few French turns of phrase. His voice carried above all the rest in the barroom, but in the cookhouse he spoke but little except to chatter some French with Paul Cyr or to demand abruptly, "Sugar!" "Cabbage!" or "*Patate!*" The Dutchmen all looked at him with some fear in their faces, for he wore a grim look most of the time and was obviously strong enough to destroy any two of them in a fit of anger. His name was Robichaud and he was known as Robbie.

He was given to playing cards every night if he could find mates enough to stay awake with him. He played with a sort of ferocity; he was not to be diverted from the cards and grew immediately irritated if someone at the table spoke of any other subject but the poker game. As for complaints, fast shuffles, or even a hint of dishonesty, Robbie would tolerate none of them. One evening, when one of the "local" lumberjacks, a Yankee-type who made his home in Oquossoc, started an argument over the proper cutting of the deck, Robbie lifted one large foot, placed it against the man's chair, and sent chair and man both spinning off against the bunks. He did not speak at all, nor even look up to see where the man had fallen. He hitched his own chair over to fill up the space just made vacant and went on dealing.

At piling his wood, however, Robbie was obviously given to ways that were dark. Jim Kidder bent down before one pile that had not yet been scaled and pulled out a stick that was barely a foot long — just a butt end that had been painstakingly fitted in among the four-foot sticks to look as if it reached all the way through. Instead, it merely concealed the fact that the wood had been carefully piled over a rock. And the woodpile itself, instead of being arranged so that a man with a stamp axe could mark both ends of every stick according to regulations, had been butted against a high rock so you could not walk behind it. It was impossible, therefore, to see how much of the pile was false front. Kidder, after tossing the short stick aside, pulled a few bolts off the top of the pile, then gave the whole pile a downward wave of his hand.

"Pull the bastard down!" he told us. Whereupon Harold and I began to remove stick after stick and toss the wood behind us into a loose heap, while Jim Kidder pulled down the other end. Finally, with the wood all jumbled on the snow, Jim attacked the other piles. He borrowed my stamp axe to search out short sticks that would shrink into the pile at a blow. He found one or two in every pile, and each time consigned the whole pile to demolition, so that within an hour all the handsomely stacked wood had been scattered in loose heaps over the clearing. Jim Kidder picked up the short sticks and fired them into the brush.

"Who's *this* sonofabitching con man?" he demanded. "One of those lulus from Lunenburg County?"

"This is Robichaud," I said. "Big Frenchman from Rumford."

"Well, you tell the prick he can pull that sleight of hand down there in the mill yard but if he tries to give me a short scale I'll boot his fat French ass from here to Magalloway. We count twelve inches to the foot up here. Right, Harold?"

Harold agreed that was right.

"And, Bob," said Jim, "you can see them sticks are stamped on both ends. And make him set the piles out where somebody besides a hedgehog can crawl behind them."

I said I would. But I was not at all sure a simple word from me was going to lead big Robbie into ways of righteousness. Robbie, however, got word of his trouble before noon was high and he carried a smoking temper back to camp.

Robbie had not seen Jim Kidder come in and did not notice him as Jim sat at our end of the long table at dinnertime. Jim had placed himself between Harold and me, on the long wooden bench, hunched down over his plate, eating fried potatoes and seeming to pay little mind to the clatter around him. Robbie strode into the cookshack accompanied by three or four of the Dutchmen, who seemed at pains to provide him plenty of seaway. Robbie did not take his seat at once but clumped clear across the room to fetch cigarettes from a jacket he had hung by the far door. At the head of our table, he paused to glare down at Tim, who was watching him wide-eyed.

"I find that cocksucking head scaler, by the jumped-up Jesus, I shove his fucking scale rule two miles up his fucking ass!" Robbie declared. "Pullen my piles down! I'll pile him, the prick! He better . . ."

Jim Kidder had somehow slid his long legs out from under the table with hardly a stir and seemed to glide like a dancer up the whole length of the table. The cook, who saw him coming, jumped behind the stove. Robbie saw him but apparently had no notion of who he might be.

"You found him!" Jim Kidder roared. And as fast as thought he brought one of his tremendous hands up in a half circle and laid it with all his force on Robbie's face. The big man was lifted clear off the floor by the blow, which sounded exactly as if someone had brought an axe full-strength down on a butt of beef. In an instant the round bald head and the flushed angry face had disappeared from our view. Everyone seemed to rise up at once, knocking pans and pannikins over and sending a swift finger of milk down along the oilcloth. Robbie lay full-length on the floor, his head partly under the stove. It seemed a miracle he had not knocked the firebox skittering and started a major disaster. For half a breath I wondered if the man were dead. But he worked his head out from under the stove and sat there dazed, one hand to his jaw. Blood was running in a small stream from a corner of his mouth.

"I don't answer to none of those names!" Kidder told him, in a voice slightly smaller than a shout. "And any time you want to start shoving that scale rule, you step right up!"

"I had enough," Robbie muttered. And the room was so utterly still that everyone in the place heard every word. We could hear the breath snorting from Kidder's nostrils still. But Jim turned to look at us all and spoke in a remarkably quiet tone.

"Anybody else want to dispute the scale?" he inquired. "While I got the chest open?"

No one did and Jim bent over to examine Robbie's face. He looked up at the cook. "Better get him a cloth to wipe that blood," he said. The cook jumped to tear off a length of lint, as if an alarm bell had rung. Robbie, sitting erect now, glumly accepted the cloth from the cook and dabbed it at his mouth, noting most unhappily how quickly the fresh blood soaked the cloth through. When he was able to get to his feet he moved slowly back to his bunk and did not return to get his dinner.

"Guess you'll have to chew his meat for him ahead of time," said Kidder. Several men laughed then. But most of the woodsmen still sat aghast, some watching Kidder as if he might still decide to bring the whole shack down around their shoulders, and a few staring after Robbie, who was lost now in the dark depths of the barroom.

The sun was gone when the pung came back to take

Kidder away but it was light enough still so the lantern dangling beneath the pung hardly seemed to glow. Jim Kidder, all jollity again, banged both Harold and me on the shoulders as the pung drew close.

"Stay offa that high wine," he charged us. "And don't get to arguing over who takes the top bunk. Make sure them seams are caulked!"

The driver of the pung brought his horse up to us at a walk and began to call to us while he was still yards away.

"Christ all Jesus!" he hailed us. "What sort of a critter did you wish on me for a passenger? Why you know what that son of a bitch done out there to the depot? Whoa back now! Whoa!"

As he brought his steaming horse to a stop close by, we all admitted interest in whatever wild deed cook or cookee might have been party to at the railroad station. The driver pushed his wool stocking cap back from his brow and sent a small spurt of tobacco juice into the snow.

"He some kind of a half-assed cop, or what? That young feller?"

Jim Kidder boomed out his laugh. He turned his beaming burnished face to mine.

"He suffers from dee-*lusions*. Right, Bob?"

"Well he sure as hell suffers from something. Serious too. You know, he and the old feller never spoke a word all the way. To each other, I mean. The old man bent my ear about what a hell of a man he'd been to fight in his young days, how he'd have beat the shit out of the whole campload if they'd used him as bad as he'd been used here. God, he like to set my teeth to chattering, I took such a fright. But when they got there to the platform, the young feller offered me a dollar and course I said there weren't no charge and he muttered some fucking thing about it being government expense. I didn't pay it no mind at the time. But when I got my sled steel loaded, the two of them was getting their luggage up the car steps. They just about made her too before the Old Cannonball started to move. There was the old guy partway up the steps and the young feller behind him. And the young feller he grabbed the old guy right up under the armpit and like to pitched him headfirst right up the steps. 'Come on now!' he says. 'I don't want I should have to handcuff you again!' *Again*, for Christ's sake! Like he'd had him handcuffed all the way down! Well, the old feller he stumbled up on the platform and he turned and looked at the young feller as if the kid was crazy. He never said a fucking word. He just stared at that kid like he seen him turn himself inside out. If I ever see a guy scared shitless! . . . By now the train was pulling away and the young feller he looks back at me and yells. 'Good work there! I'll git him back safe now!' Good work. Well, shit a goddamn! Like we'd captured a crook together!"

"Maybe he'll cut you in on the reward!" Kidder laughed. He was climbing up beside the driver now, and grinning down at us. "Never knew you was harboring such a dangerous character, did you?"

"What I don't understand," said the driver, shaking out the reins, "is what the hell he stood to gain by putting any such an act on for my benefit. Less'n it was on the level."

"Oh, he probably saw you one day with that tin badge you wear for being a deputy fire warden," said Kidder. "He figured you was a fellow lawman, that's all."

"Oh lawman balls! Come *on* now, Bess! Get up in there!" The driver looked back over his shoulder at Harold and me. "What the hell was he? Wangan clerk here, or what?"

Harold and I replied in unison.

"Cookee!"

"Cookee! Well, shit a goddamn!"

Now of course the camp had no cookee at all. It was the

cook himself, in a scratchy voice that hardly carried over the yard, who cried out the meal calls and bid us turn out before dawn. And because I was one man whose daily duties were not completely defined, much of the cookeeing was left to me, despite my earning half again as much as a cookee was drawing. With Harold's help I guided the crosscut back and forth over the wood in the dooryard to junk up the long hardwood logs into stove lengths. Then I split it and stacked it under the roof of the small shed where the molasses barrel was kept. Working about the kitchen this way I developed a close acquaintance with the raw materials that went into the products that came out of the cook's pots and pans and in this way acquired a permanent distaste for some of the regular dishes. But I made fast friends with the cook, even though he had a sort of superstition about my carrying my stamp axe into the cookshack. He would lift one hand and wave it frantically when he saw me about to set the axe down by the wall.

"Get that thing out of here for God's sake! I don't *like* that thing!"

It seemed an innocuous tool to me, somewhat like a small pickax, with one point blunted and fashioned into a square brand. The sharp end of the axe was for snagging hold of a pulp stick and pulling it into position. The cook seemed obsessed with the notion that he might step on the axe someday and flip it up to make a hole in his carcass.

To express his gratitude to me for "choring him up" with buckets of lake water and spring water, and with bolts of maple and birch for his woodbox, he would set out plates of molasses cookies in the middle of the forenoon for Harold and me to take with our tea. Ordinarily, like most cooks, he would fly into a temper to see outside men wandering into his precincts in search of food during off-hours. But he made me one of his team and welcomed the scaler as my sidekick, no matter what time of day we might seek morsels of grub to stay us. The truth was, and the cook never knew it, that neither Harold nor I ever ate a complete meal from his table, except when sheer hunger forced us to choke down some of the leathery bacon to go with our doughnuts and tea at breakfast. The new cook's doughnuts were no different from the old one's, they being well saturated with used fat and quick to grow stale and hard as small knots of wood. But softened in hot tea, they would still the complaints of our ravening young stomachs for an hour or two. The molasses cookies, however, were choice as a fond mother's best. And we always received them fresh and sometimes still warm from the oven.

My cookee chores led me one day — while the McCormicks were still sending "downriver" from time to time in hopes of bringing a new cookee in — to carry a long-handled shovel over to dig out the spring so a bucket could be sunk into it over the edge. The spring ran out of a bank close by the camp where Wallace and his wife holed up, and it ran into a sandy pool that filled up from time to time with silt. Wallace's camp was shaped like a small way station, a one-story oblong building with square windows in either end and two on each long side, a board door hung on leather hinges, with a doorstep of a single length of hand-hewn hemlock.

When Wallace and I walked there together this day our feet were hushed by the snow, and Mrs. McCormick had no notice of our coming until Wallace spoke to me, just a few yards from her window. Then she opened the door quickly and looked out at us, a very plump woman, with a round, spectacled face and her hair braided about her crown, to add to the roundness of her whole head. She was a perfect picture of an old-fashioned schoolmarm, which is just what she had been when Wallace found her while she was entering her forties and wooed her away.

53

"I was sleeping," she announced, not in a complaining way but just to explain, I suppose, the lack of activity about her house. Of course it was not a house but hardly more than a shanty, covered all over, roof and walls, with lengths of black tar paper, strapped by long sticks and lengths of scantling, such as had been left over from sawing out boards for floors and roofs.

"I thought we heard somebody snoring," said Wallace, in such a tender way that I was startled. It had never seemed to me that a man so notoriously grasping and so given to nursing grudges against all his fellows could ever experience, much less exhibit, any gentle emotion at all. His wife, to whom he had been married now about a dozen years, giggled like a girl. "I wasn't *snoring*," she protested and bent on Wallace a look so fond that it almost softened me toward the man. Forever afterward I dwelt on the devotion that these two shared for one another, unwilling at first to believe it could exist at all and finally I suppose granting to my better self that there might possibly be some atom of lovability in even the nastiest man alive.

Wallace was not the nastiest man alive, even in our world. Most of us would have awarded that title to Tim, who was a more sophisticated man, he having traveled often to Boston, and other large cities, and having developed more complex means of expressing his distaste for all the rest of the world's people. It was traditional in that neighborhood to refer to men who hailed from the Magalloway district — just across the line in New Hampshire — as "Magalloway skunks." It was a sobriquet that the skunks themselves accepted with wry resignation, and even a sort of pride. But Tim applied this term to every human creature who dwelt anywhere in the surrounding counties, not one of whom, to hear him tell it, had ever really owned clear title to a pot to piss in or a window to throw it out of. There were not only Oquossoc skunks, in Tim's lexicon, but Rangeley skunks and Rumford skunks and Farmington skunks as well as skunks from all the townships in between. There was always implicit in this wholesale generic condemnation the faith that the McCormicks themselves owned citizenship in some land where dwelt only sober, industrious men who paid their debts, cheerfully worked long hours for low wages, banked all their pennies, and habitually yielded up to the McCormicks the long end of every bargain. When Tim once, in my hearing, undertook to lay a compliment upon a man named Corrow, whose home brew we were freely guzzling, he unctuously offered that he had "never heard a man say Fred *Corrow* owed him any money." It was the highest praise Tim knew how to express. It went without saying, among those acquainted with Tim, that this same Corrow would return to his niche among the "Oquossoc skunks" once Tim had left his rooftree. For there was not a man living for whom Tim would speak a good word beyond the man's hearing.

The fact that they were heartily disliked throughout the woodlands was never lost on the McCormicks. Once, in an angry discussion with me, Wallace laid their unpopularity to the fact that the "*fine* people hereabouts" were all jealous because the brothers could take a contract to land ten thousand cords on the lake in ledgy woebegone country and make money, where the local "skunks" would "go in and *starve*."

It did not strike me that there was much about the McCormicks to be jealous of, for their lives seemed singularly joyless. Their vacations were usually spent touring, all three of them, through some of the cities in Canada, where, Tim reported, the people "always had their hands out for *tips*." And I never heard either of them talk of these trips

55

with anything but scorn for the folk they met and outrage at the cost. Once, it is true, I heard Wallace report happily on a one-day jaunt he made down the lake to a summer resort, now closed for the winter, to take Thanksgiving dinner with the owner — a man very nearly as mean as the McCormicks themselves and one equally given to mortifying his own flesh that his bank account might prosper. "Well," said Wallace smugly, as he walked up from the dock that evening, "I got my dung bag all full of good sauerkraut!" Sauerkraut for Thanksgiving! Jesus, I told myself, what a hell of a thing to rejoice about.

Nor did it seem to me that there were many grown-ups in the land who would begrudge Wallace and his wife anything about the life they led in that little board shack, with just a thickness of tar paper to shut out the near-arctic cold, a toilet in the woods, water fetched in a pail, a tiny range to cook on, a small box stove for heat, and a chamber pot for company through the night, with fresh meat brought in but twice a month and that not often hearty enough to make a solid meal on. It is true that Wallace and his wife lived better than we or the woodsmen did, for they did receive somewhat fancier canned goods and Mrs. McCormick made her own pastries.

Tim, of course, like the woodsmen, must have found little comfort in his meals, which were so often the chief solace of most working men of that day. God knows our table little resembled the legends of lumberjack fare on which I had been raised: fragrant rich coffee and frying home-cured bacon, and beans devotedly baked in great crocks in the ground. Our lumberjacks ate beans, but they were always soldier beans, baked without molasses into pale mush in shallow oblong tins — to the very consistency, one Dutchman glumly observed one day, of shit. As for the rest of our diet, it had improved somewhat with the advent of the new cook, but there was still no juice in the meat and very little in the turnips. The desserts were usually pies filled with some patent gelatin dug out of wooden pails and packed untreated into pastry shells to be baked into pies in the oven. Once I had been made acquainted with these pails of pie-filling labeled "Artificial Lemon" and "Artificial Raspberry," I vowed I would eat no more of them while the world should stand, even though, like every child brought up poor in Boston, I had been addicted all my life to "sweet stuff" after supper.

Most of the workmen simply despised the two brothers for the reason that a dog hates the hand that holds the whip. And there were many people in the countryside round-about who had been beat by the McCormicks in a deal, or who had forgotten to agree on a wage before going to work for them and found themselves paid off at a dollar a day less than they had counted on. So it was not jealousy at all that moved most people who spoke ill of the brothers but honest resentment.

There was one man who did manage to beat the Mc-Cormicks in the strangest way — without ever feeling, I am sure, that he had got any more out of them than his minimal deserts. He was the cookee who was shipped up finally from Rumford, a gnarled little man named Joe Arsenault, who could have been either fifty or eighty years old and would not say which.

Joe was reputedly one of the best splitters of firewood in the whole back end of the state, a man who could take a yard full of stove-length chunks of wood other men had given up on and split it all into neat sticks ready for the fire. His secret, he told me once, was to take a stick full of knots and clout it right in the center of the toughest knot — as a general might choose to attack a fortified position right where the enemy had massed his greatest forces — the theory being I suppose that this was, essentially, the weakest spot, or the keystone that, once dislodged, would bring the whole contraption tumbling. Joe did, in my sight, often crack open an impossibly gnarly old butt end until only a few twisted fibers held it together.

The grim joy he took in his work must have been rooted in his satisfaction at continually accomplishing the impossible. In his first few days on the job he had the woodbox overflowing and the shed almost stacked full of good clean hardwood that would last the cook a month. The McCormicks, rejoicing over the fact that, at a dollar a day, they were getting some three dollars' worth of labor out of the old man, took to joshing him pleasantly whenever he crossed their vision and to making frequent note of the devout way he put in endless hours on the woodpile. Indeed, it did seem that splitting wood was Joe's recreation, which he turned to whenever he could win time away from his regular chores about the cookshack. He was up before the cook, woke up the camp in a voice like that of a lonesome wolf, and moved solemnly and silently up and down the long tables to keep the pannikins filled and keep the tea and coffee flowing. Joe was not a talkative man but he was consistently friendly, never out of temper, and never, under any circumstances, given to complaining, as some cookees had, of an overload of work.

But one Sunday, as I watched Joe through our cabin window deftly splitting a fat length of cedar into tiny sticks of kindling, I saw him halt his axe almost in midflight, then set it gently down upon the split wood. He remained slightly bent over, as he had been when completing his stroke with the axe. Then, very painfully, he set both his hands at the small of his back. I had seen old men make this very move a dozen times in my life and had heard them complain of a "dropped stitch" or of being "taken" suddenly in the region of the kidneys by a mysterious ache. Indeed, Joe at that moment might have posed for one of the familiar ads for Doan's Kidney Pills. But Joe remained in that pose an inordinate length of time, as if the world had stopped and he had been turned to stone. He did not so much as move his head. Then, just as I was about to run down and see if he had indeed fallen under some magic spell, Joe unbent a tiny degree and instantly froze rigid again, this time with his head up and his back very slightly unbent. In this position, and still holding both hands clamped above his kidneys, he shuffled through the dirty snow, in and out among the tumbled bolts of firewood, back inside the cookshack.

And that is where Joe stayed. When I went down a few minutes later, Joe sat on his bunk, with his heavy jacket off now but still wearing his hat. He kept rubbing his back with his left hand. His expression was utterly calm, actually inscrutable, as if whatever had befallen his body was of no great concern to his mind. He lifted his eyes at my appearance and when I asked how he felt he told me "Good." The cook suggested I bring down a bottle of Sloan's Liniment from the wangan and I hastened to fetch this ancient cure-all, which, for all I knew, would promptly set old Joe to leaping about again like a youth. Joe permitted the cook to remove his shirt and get his arms out of his dirty union suit, so it could be dropped to his waist. His skin was chalk white, mottled by irregular dark spots like moles, and stretched tight over the ribs and the knobs of his spine. The cook took a dollop of liniment into his hand and smeared it over the small of Joe's back. The piercing smell of it made us blink. It seemed as if it might start the old man to hopping from the sting. But he sat impassively as the cook patiently rubbed the liniment around and around, until all the skin in the area glowed red.

"You get those pores open," the cook explained to me, "and the stuff'll burn that soreness right out."

After the treatment was over, Joe silently worked his arms back into the underwear sleeves and slowly stood erect. He remained on his feet, slightly bent, for a few seconds, winced slightly, shook his head, and sat right down again in a straight chair beside his bunk.

"No better?" the cook asked him.

"Nope," said Joe.

"Give it time to work. It takes time."

It took very nearly forever with Joe. For Joe remained in that chair, when he was not in bed, or toiling painfully to the outhouse, for the next two weeks. He did no more work about the kitchen. He split no more wood. He carried no more pails of water from the spring. He no longer sent his wailing call out over the dooryard in the dark of the morning. Instead, he arose each morning, shuffled out to urinate outside the door, washed himself about the hands and face, then sat down and by the light of a small kerosene lamp read stories from the suitcase full of magazines (he called them books) that he had toted in with him. He had uncounted copies of *Adventure*, *Ace-High*, *Argosy*, *Blue Book*, and other ancient pulp magazines whose names I have forgotten. Some of them, from their dog-eared state, must have followed Joe for about five years. But Joe immersed himself so deeply in them still that not all the clatter of meals being set out and done away with could cause him to raise his eyes from his "book." Tim and Wallace would inquire solicitously from time to time if his back was any better and Joe would tell them "Nope," and go back to his reading. Then they began to ask him instead when he thought he might feel like getting back to work, tomorrow perhaps? Or next week? And Joe would pull himself out of the faraway world the story had carried him to, would meet their glance with his own mild eye, and tell them, "Don' know."

The McCormicks were always especially sensitive to the possibility of legal action against them for some injury on the job. They carried no insurance but they did have a deal with the doctor in Rangeley to minimize their risks. They would hustle an injured man promptly over to the doctor and the doctor would belittle the injury as much as possible. When the Dutchman broke his leg it was the doctor who assured him he could travel home in safety, to get himself as far away as could be from the McCormicks' jurisdiction. And if any man visited the doctor on his own, for treatment of hemorrhoids, a cracked finger, a persistent bellyache, or an itch, the McCormicks would get a full report at once, no matter if the man had been simply fretting over the consequence of a visit to one of the semiprofessional whores in Oquossoc.

So they were not about to stir up any resentment in old Joe's breast, lest he be prompted to inquire into his legal remedies. Still, there was no room on their payroll for two cookees. If the truth had been acknowledged, they could have kept Joe in idleness for a month and still have been ahead of the game. But the McCormicks could not adjust their temperaments to any such reckoning. At first they must have felt that a few days off for Joe would hardly cost them at all, he having done more than a good month's work in a week. But little by little they saw their end of the deal growing shorter and shorter. While they were still reluctant to let go of their prize, they could foresee a time when they might have hold of the losing end, and they fretted by night, with Wallace squatted by his blowing gas lantern, over how long it was proper for a man to expect to get his chuck for nothing.

The day came at last, after Joe had devoted himself to

reading for two days past a full two weeks, when Tim and Wallace felt they would have to accept the risk of a lawsuit and move old Joe out.

For all the fretting and muttering the brothers had done beforehand, their firing of Joe Arsenault proved simple indeed. Wallace, after the cookshack had emptied of all but the cook and Harold and me — and of course Joe himself, impassively reading — explained to Joe in a tone even quieter than the one he normally used that they had to bring in a new cookee and they'd have no place to stow him unless they took Joe out. Joe would be better off, Wallace urged him, where he'd have someone to look out for him, and a doctor nearby if his condition grew worse. No doubt about it, Wallace declared, this must be some ailment of long standing and all the work could possibly have done was start it up again.

Joe listened to all this politely, his book closed over one finger. And when Wallace had finished, Joe said, just as quietly, "Aw right. I go home tomorrow." Whereupon he returned to his story and did not even lift his head again to make note as Wallace left the camp.

That evening Wallace and Tim rejoiced together over the ease with which they had got this burden off their backs. And Wallace, in figuring out Joe's pay, was even exhilarated enough to dare to charge Joe for the food he had consumed during his "laying by" time. He subtracted fifty cents a day from the small sum Joe had coming, then turned to Harold and me, surprisingly prim in his wire-framed reading glasses, and spoke in a tiny, questioning tone:

"A man shouldn't get his grub *free* in the woods. When he's not working?"

Harold and I, not ready either to agree or argue, merely shook our heads silently. Wallace, however, had already answered himself. He nodded, and handed me the slip, so that I could make out the company order.

Joe's departure next day was all goodwill and heartiness, with even Tim taking his own turn at shaking Joe's hand and warning him to stay away from the Rumford girls until his back was better. Joe smiled at everybody and shook every offered hand. There had been no such farewell for anyone else in all my time there, not even for the men who had been going back to Nova Scotia to stay.

The joy, of course, was exuded largely by the McCormicks themselves, who infected all the lesser folk with their own blessed relief at having extricated themselves so painlessly from what might have been a desperate fix. But watching the benign, inwardly smiling expression Joe Arsenault wore as he saw his two straw suitcases, swollen with his gear and his precious magazines, stowed on the sled, Harold and I began to wonder and later to ask each other aloud if the McCormicks had indeed beaten old Joe in the deal. What if the old man had simply conspired with himself to win two weeks of quiet in the woods, to read his stories and invite his weary soul? What else would have prompted him to tote such a load of printed matter into camp — enough to have kept him occupied through a whole winter and a half of spare time? For a cookee ordinarily worked from before dawn to long past dark, seven whole days of every mortal week.

Neither of us, it was true, would have rated two weeks' confinement in a camp the McCormicks operated any higher than a short term in Farmington jail. But for old Joe, whose days otherwise might have been spent in somebody's attic, or the county home, a woods camp might have seemed like a holy refuge. The cookee, after all, like the cook, had a bunk of his own, built to snuggle close to the stovepipe; and of course, like the cook, he had the pick of the pot. Had there been any small portion of the common fare that deserved to be named a tidbit, Joe and the cook would have divided it. And there was the faithful stove, the snug bed, and the long, quiet daylight hours when a man could sink himself over the ears in the enfolding world of fiction, where men of wit and daring drew thunderbolts from their holsters with the speed of light to visit doom on mustached villains, coped with impassable mountains, blinding blizzards, or storms at sea, helped the sore beset, worked miracles with a suddenness that passed all understanding, escaped alive from hopeless battles, from shipwrecks, stampedes, stark hunger, raging cannibals, blood-maddened sharks, pits of fire, foreign jails, howling Indians, outlaws, or creatures from outside the world, and earned fortune, preferment, and the love of fair ladies.

For all we knew Joe's whole winters were spent in just this fashion — splitting tough chunks of firewood with fierce devotion for two weeks, then settling in to imbibe his beloved stories for as long as the boss could bear him. Harold and I used to laugh long and loud when we thought of how Joe must be moving from camp to camp, toting his library with him, to earn himself long weeks of comfort and to go back to Rumford finally with tobacco money jingling in his pocket.

There was damn little else to laugh about in camp, except our own discomforts. We laughed as we saw ourselves one night, the two of us bent over a deep hole in the ice, rinsing our newly boiled underclothes in the bitter cold lake water, then wringing them out, one at either end of a dripping cruller of clothing, with our fingers near ready to drop off from the chill. But we at least had a chance to boil up our clothes with soap in a bucket. We had no body lice or bedbugs to set our flesh on fire. We had cotton-filled mattresses rather than straw to rest our frames on, and we each had a whole bunk of his own, without a restless bunkmate to kick us awake or snore into our ears.

In the barroom, the woodsmen owned no such comforts. They sank into black sleep from desperate weariness, escaping their hunger and their other aches by courting unconsciousness soon after they had swallowed their supper. A few of course always tried to rescue some moments out of the day by sitting up at cards or by trying to argue in their high-pitched voices matters of religion and morals.

The loudest voice in the barroom, as always, belonged to Robbie, who despite his defeat by Jim Kidder still held the whole company of innocent Dutchmen in a state of awe. He was big and strong and fierce-looking, and talked so loud that no two voices could prevail against his. Many of the Dutchmen listened to him almost in terror of the blasphemy he uttered and the villainous oaths he used to underline all his discourse. They were moved on occasion to dispute him mildly, only to discover that his depravity reached far deeper than they dreamed. For not only did he take God's name in vain through the most revoltingly obscene combinations of phrases, he even laughed bitterly at the very suggestion that there was ever any Supreme Being at all. Religion, he told his mesmerized listeners, was the greatest fucking fraud in the world and only fools were its victims. As for him, he feared no hell and yearned for no paradise beyond the fleshpots of Lewiston or Rumford Falls.

No one ever spoke of attending church on Sunday. Had anyone been so driven by devotion as to trek the five miles to Oquossoc, he'd have found only the tiny Catholic church functioning there, and no Protestant church closer than Rangeley, seven miles farther east. Sundays in our camp were given over partly to stamping the piled wood or patching up the swamping of the roads, or to filing and setting saws, grinding axes, delousing clothes, and cutting hair or

65

whiskers. In the time before supper, a few of the young Dutchmen might forgather in the office camp, to buy apples or candy bars and consume them in our company and, if the McCormicks were out of earshot, to gripe about the dreary diet, the stingy wage, the dollar-a-day fee for board, and the shoddy pants our wangan supplied. More than one young man went out to work with his raw underwear showing through his pants, and sometimes the men would sit with us and try to patch their clothes with precious safety pins. And they would take us into their arguments over the rights and wrongs of the world, and report, in tones of half-disbelieving alarm, that Robbie had revealed himself as an *atheist*.

"He don't even believe in a future life!" Will Ernst told me and watched to see my eyes grow wide.

"*I* think there's a life after death," said Nathan Aulenbach, in his little-boy voice. "I think there *has* to be!"

※ 3 ※

THE feeder at the McCormick Brothers camp was a
spindly fellow named Gus Lorry, a man in his forties
who always put me in mind of an insect, for his legs and
arms seemed jointed like a grasshopper's. He and Paul Cyr
were the only men in camp who let their beards grow full,
but Gus never combed his, nor combed his hair either, so
black wiry strands stuck this way and that all around his
head and face.

It was Gus's job to feed the horses, to keep their quarters
clean, and occasionally to hitch one to a bunk and take him
into the woods to haul in a few logs for firewood. He also
curried the creatures, knew how to mend harness, and could
fashion a war bridle to keep a horse from cavorting too
wildly when there was need to do some small surgery on
him. Yet he seemed to hate horses, to distrust their motives
and belittle their good sense.

"You can talk all you mind to," he used to say. "But they
was never a hoss foaled could outpull or outsmart
a ox."

According to Gus's story, his glory days had been spent

along the Swift Diamond, where he drove a team of oxen on long-lumber jobs past counting. Gus was an excitable man, who would begin to wave his arms and take his breath in gasps in response to even the mildest impulse. When he would talk of the perils and satisfactions of teaming a yoke of oxen over snow and ice and corduroy with a load of sixty-foot logs, he would begin to leap and shout as if he once more held the goad stick in his hands.

"Gee! You son of a bitch!" he would scream, and would jab an invisible flank with an imaginary goad, fiercely enough to make a listener flinch. Then, lost in his vision, he would lurch and shout and fight for his balance as the tortured beasts bent madly to their jobs and the careening legs actually smoked from the friction.

The McCormicks, who belittled every man they knew, insisted out of Gus's hearing that he had never held a job in his life more demanding than shoveling horseshit and tossing flakes of hay into a manger. But Gus continued to comport himself like a teamster, especially when he took the twitch horse out for firewood. The hillside then would ring with his curses as he drove the stumbling horse over stumps and snowdrifts, pretending that he was in imminent peril of being overrun by the load of logs and carried, horse and all, over a precipice.

Gus would yell most wildly, however, when he had to deal with Danny, the meanest horse in the hovel, a knot-headed young gelding who seemed to find a perverse delight in terrifying any newcomer who came within range of his rolling eye. Often, loafing through the early evening in the office camp, we would all sit up straight to attend the sudden thundering of iron shoes on the hovel floor, and the wild banging as Danny strained against his halter and let his hind hooves fly. Then Gus Lorry's scream would sound over all the turmoil: "Whoa! You son of a whore! Whoa!

You cocksucker! I'll beat out your fucking brains with this pitchfork. Whoa! WHOA!"

We all laughed then, for we knew it was Danny having his fun. Danny disliked nearly every man alive but he had developed a special distaste for Gus, who had reacted with a leap into the air when Danny first began to rear at his presence. As for me, I had welcomed Gus with a special warmth, because until he came to camp — just before the lake froze over — I had had to fill in on feeding, watering, and bedding down the sixteen bony, cross-grained, hungry-gutted horses who had been carried in by barge from their frozen pasture.

Some years earlier, when, as a small boy on a New Hampshire farm, I had been awarded the barn chores to do every morning, I had learned to cater to the most evil-tempered animal in the lot (there seemed to be one in every barn) by taking special care of his cravings. And so I had set out immediately here to make a friend of Danny. There were always some half-spoiled apples in the wangan that I could stuff in my pockets to feed to Danny and once or twice I saved him a bit of candy bar. Although he had first greeted me, as he greeted everyone, by trying to kick a few poles loose from his stall, he began at last to let me sidle in to bring him his daily treat. He would lay his long ears back and his eyes would show white as he caught sight of me behind his stall, but after the first three or four spasms of anger, he kept his temper in check. Eventually I could even stroke his neck — a gesture that earlier might have driven him into a murderous frenzy — and could take the metal currycomb to his gaunt ribs without danger of his trying to rip free of his halter to destroy me.

After Gus came, I still was needed on occasion to perform some chore for Danny. But I was excused from filling the cribs with hay and shoveling out the accumulated droppings and from toting endless buckets of water to fill those sixteen bottomless bellies. So Gus, for all his scruffiness and general no-account air, remained a friend of mine. At least I was grateful, every day, that he had come.

It was obvious, however, that Gus irritated the McCormicks beyond reason. Had he been content to skulk about the hovel and barroom and otherwise keep his distance, as the woodsmen did, no doubt Tim and Wallace would have borne his presence with a modicum of patience, in view of the fact that he was costing them but a dollar a day. But Gus was forever bursting into the office camp with some tale of impending disaster in the hovel, some news of broken bunk chains or a missing hayfork or a horse that seemed like to throw a shoe. Like every event in Gus's life, those petty mishaps made his eyes to pop and his breath to rasp in his throat, as if he had run all the way to get help before the barn burned down. After ten or twelve such emergency visits, Tim, who was usually the only boss on hand when Gus burst in, took to ignoring Gus altogether as if he was not even supposed to hear what Gus was telling. So poor Gus would stand open-mouthed, with his eyes full of pain, like a boy who had given a stupid answer in a schoolroom. One Sunday, when Gus had come panting in to report signs of colic in a chestnut mare, Tim had merely looked at him owl-eyed and then turned his back on Gus and sat down to read a torn copy of the Boston *Post*, leaving Gus forlorn. Harold and I, moved at the same instant by a surge of sympathy for poor foolish Gus, opened our mouths together to beg for the details, then laughed at each other as our words came out almost in unison. Gus, after one final glum glance at Tim, turned eagerly to describe to us the spasmodic lifting and stamping of the mare's hind feet, which he illustrated by a thumping of his own left foot on the cabin floor — a move that caused Tim to lower his paper and roll his eyes at us all.

"Let's take a look," said Harold, heaving himself up from his bunk. We trooped all three together to the hovel, with Gus muttering the whole way in a doomsday tone about the perils of leaving colic untreated, and of generally ignoring the ills of horses.

"You take it in the fall of the year," he told us, as if he were justifying himself to Tim's face, "before you start to grain 'em, a hoss can get off his feed like *that*." He gestured with one tattered mitten to indicate a snap of the fingers. "By Jesus, it don't pay to be too goddamn high and mighty when you got a sick *hoss!* You wait too long and, by Jesus, you'll have a *dead* hoss."

The chestnut mare, however, was not even close to death. She stood at ease in her stall, turning her head and slanting her ears forward to greet us and shifting from one large hoof to the other in perfect comfort, with never a stamp or a spasm.

"Wait a bit, now, and you'll see her bring that old hind leg up," Gus assured us. "Jesus, she like to gut herself back there a bit."

But the mare felt no urge to gut herself or to raise her hoof at all.

"Just like a Jesusly hoss," Gus declared. "They's worse than a fuckin child." He stood there with his mouth pulled down, seeming ready to fetch the mare a whack across the rump to force her to produce a symptom. But the mare went back to snuffling about her trough after what wisps of wiry hay still lay hidden there, and paid us no further mind.

"I God," said Gus wretchedly, "*I* don't know what ails

the critter! She like to curl up with cramps not ten minutes ago."

"They can get well in a hurry," Harold offered feebly. Gus shook his head and spat a stream of tobacco juice into the straw.

"Well, by Jesus, she *was* as colicky as I ever knowed a hoss to be. She was jerking that hind leg up like she been bit. Son of a bitch if she warn't! Jerking and stamping." Here Gus again provided his own illustration of the horse's symptoms by stamping his foot heartily on the pole floor, making the whole hovel shake. The chestnut mare turned to look at Gus with the same dubious expression Tim McCormick had bent upon him. Two other horses nickered uneasily, while Harold and I exchanged a solemn look to keep each other from laughing.

"I thought I noticed her doing that this morning," I lied. "Stamping her hind feet. I didn't know it meant anything."

Gus's face opened wide with gladness.

"Son of a bitch!" he shouted. "You see what I mean? More like a sort of spasm, warn't it?" He illustrated his meaning by jerking his own knee suddenly up to his belly.

"Kind of like that," I said.

"Colic!" Gus exclaimed. "Colic surer than hell! I seen it a hundred times!"

"Well, better keep watch of her," said Harold.

"Goddamn right," said Gus, fiercely, as if he meant to mount guard day and night.

From that time forth, Gus looked upon me as his close ally against all the world — most of which, by his accounting, stood ready to do him down in a dozen ways. He would seek me out whenever he needed a hand in getting a recalcitrant log down the hillside into camp, or had broken a bunk chain and needed moral support in bringing word of the disaster to Tim. More than that, he entrusted me with secret errands that, had I shared them with the camp, would have unmanned him.

You must understand that the image Gus tried to offer to his immediate world was that of a man it did not pay to trifle with. "I'm full of bunk chains and dynamite!" he would often declare, in voicing a defiance to some challenge that had not yet been hurled. It was therefore an act of ultimate intimacy when he approached me one morning with one hand buried inside his shirt and, after assuring himself I was alone in the office camp, drew out a talcum powder can and pushed it into my hand.

"Stick that in your budge," he muttered. "And get me some of that when you go to town."

I glanced at it quickly before I hid it away inside my own shirt. It was a purple can with a picture of a lady in a filmy gown from whom tiny flowers dripped in a shower. In large white letters on the can, the trade name was printed: ADOR-ME TALCUM. I can remember telling myself that I would *have* to share this with *somebody*. Ador-me, for Christ's sake! Hairy, grimy old Gus, reeking of horse piss, with shit in the cracks of his boots and most likely between his toes as well, about to make himself adorable by means of what we then called a French bath — a generous dosing of scented powder! But of course I could never betray Gus to anyone in the camp. It was not until months later that I told this tale to Jim Kidder and we laughed together.

At this moment, however, there was no laughing. Solemn as a secret agent, Gus looked me straight in the eye to make sure I understood the gravity of the matter. And when he saw me bury the can deep within my own shirt, he handed me a fifty-cent piece, nodded, and hurried back to the hovel.

In this period, I hiked out to Oquossoc at least once a week, to carry out mail and fetch some in, to buy odds and

73

ends of wangan that could not be secured at the storehouse — two-for-a-penny pencils, rubber bands, scratch
pads, papers of pins, and bootlaces — and occasionally to
buy a money order for one of the choppers and mail it to
his home. This transaction required me to cash a company
order first (always discounted two percent by Cliff Hill,
the thrifty gentleman who ran Mackenzie's General Store)
then to copy off on the application all the required
information, to count out the remittance, plus the fee, tuck
away the little blue receipt, and wrap the change in paper
to keep it separate from the other small funds I carried. It
was always a matter of twenty minutes or more to
complete these small offices and I never stinted them, for
the entire process required the assistance of Cliff's
sixteen-year-old daughter, a pretty girl with hair as black
as an Indian's. It was not her hair, however, but her eyes
that took my breath away. They were not brown, as they
should have been, but a deep blue. Meeting her glance
directly always sent a tingle through the pit of my
stomach, as if a small cold finger had touched me inside.
And it always happened, when I faced her through the
little post-office wicket, that she looked me directly in the
eye, quite soberly, for a full second, before she smiled, or
said my name in greeting. And it was this moment I used
to dwell on as I plodded along the empty tote road in the
early morning, through the snow-laden hardwood, toward
the highway that led to the store.

On the day I carried Gus's talcum can to town, the
weather, to use Gus's own phrase, had "moderated awfully." The tree branches were bare of snow and gleamed
wetly in the pale sun. The eaves had dripped all night. As
I set out across the lake, about noontime, the cook, emerg-
ing to draw molasses from the barrel, warned me to "watch
out for reefs" — rifts, that is, in the ice, where a warm

current might have melted most of the way through. But the ice was solid, several inches thick, heavy enough to hold a team. There was only gladness in my heart as I started across the wide white plain toward the woods on the far-off shore. Well above the point where Kennebago Stream flowed into the lake, there was an easy foot trail through the woods to the tote road that led down over the corduroy winter bridge, across the stream near Indian Rock. As I walked down this road, making the only human footprints there since the last snow, two whiskey jacks drifted, silent as falling feathers, out of the birch and popple growth and kept me company most of the way. One after another they would pace me, flying sometimes so close to my head that they might have landed on my hat brim, perching occasionally on a branch nearly level with my face, cocking a happy eye at me and uttering every now and then their low wondering whistle, as if they were expressing admiration of my clothes.

The tote road led past a cluster of solid buildings, made of varnished logs, where a rich men's fishing club had closed off much of the stream bank for themselves. Now the long asphalt shingled lodge and its huddled neighbors all wore windows that had been neatly blanked out by shutters of unpainted wood. Water still ran off a few of the shaded roofs; on one corner a few dripping icicles persisted. It was lonely as an abandoned fort.

There were no other buildings along the way until I reached the highway, where I could hear a car a long way off clanking a loose chain against its fender. The snow remained on the highway only along its edges and in shady patches, where it lay soiled with horse droppings and churned-up yellow mud. Here the village began in small square homely houses, painted white, all set well back from the road with footpaths of soiled snow or raw mud leading to the front stoops. There were perhaps six such houses, all a rod or two apart, along the highway between the tote road corner and the village proper, where the railroad station gathered all the rest of the village about it.

Mackenzie's store was the tallest and widest building of all, two stories and an attic high, a red building with two show windows filling the front and a platform running the whole width. Wooden stairs at one side led to the "rent" upstairs where Cliff Hill and his family dwelt. Because the show windows were cluttered with old pasteboard signs, two or three summers old, and faded stickers that celebrated Ward's Tip-Top Bread, Mount Zircon Ginger Ale, and Union Leader Pipe Tobacco, it was not possible in the daytime to know from any distance if the store had a light in it. I approached the place with slowing steps, for I had not yet selected the phrases I would use to make it clear, without obvious concern, that the talcum powder was not for me.

There was hardly anyone moving in the whole village, although I could hear muffled voices from inside the Dead Rat, a long shabby frame building that offered cheap room and board to lumberjacks. As I passed the front porch here, the door opened and a plump man in a checkered wool shirt called me by name. I waited in the roadway for him to come out to me. He was Jim Sullivan, a summertime guide who had often hidden his bottle in my closet at the hotel, so he could stay the onslaught of ennui when he fetched a whole string of girl campers for an outing at the hotel. While his blue-bloomered charges gorged on packaged ice cream, Coca-Cola, and chocolate bars, Jim, their "nature counselor," would wander idly off to my cell and improve his own nature with constant small applications of Old Tom gin.

"Hey!" Jim greeted me. "You got a cookee yet up there? I could use that job!"

But we had a cookee then, a pale Frenchman, or rather a man with a French name, who looked and talked like a hotel clerk but did his work with good humor and enthusiasm. Anyway, I told him, I didn't know if the McCormicks and merry Jim Sullivan were meant to dwell together.

"Christ!" said Jim. "Don't worry about me. I can get along with a sore-assed skunk. I need a place where I can stay dry for a few months."

Jim did exude an aroma of a man who had slept for several nights in a saloon. All his clothes seemed imbued with the smell of flat beer and stale cigars.

"I got to get out of *this* Jesusly little paradise," Jim went on. "A feller stay around this place any length of time and he'll get to drinking!" Jim said this with the happiest of smiles, grabbing the slack of my jacket meanwhile and shaking it affectionately. "Hey, we had some quite times at *your* little pigpen last summer!"

All the "times" I could recall were the sudden appearance of Jim's florid and beaming countenance in the hotel front office with a bottle in a brown bag, which he would entrust to my care, and his departure two hours later, with an even brighter eye and a rather fixed but nonetheless beatific smile. I agreed, however, that they had been occasions worth looking back upon.

"Say!" said Jim suddenly, pulling my jacket up tight. "You know who's back in town? Your old friend Jenny!"

Jenny had been, I was convinced, a far better friend of Jim's than she ever had been of mine, or a more intimate one. Yet I felt then, and still feel, that she was one of the nicest ladies I ever knew. Not that any of the proper people in those parts would have named her "nice." She used bad words freely and made no more than a fleeting effort to gloss the fact that, as she sometimes put it, her shoes were occasionally under some bed other than her husband's. Indeed, when I first knew her, she was acknowledged the full-time consort of the man she worked for — a skinny, white-haired man who seemed, to me at least, far too old for sex. But Jenny had no air of the sex queen about her. She was as natural as a pet cat. A small woman, with hair for that era unfashionably long and usually hanging to her shoulders, she had the appearance of an Indian. She was dark, brown-eyed, and muscular, with black hair that wore no wave or curl. She had a way of looking at you that was completely disarming — not coquettish or sly or contrived but open and friendly as the manner of a child. Her mouth was wider than most women's and her lips full, nearly always smiling, or seeming just about to smile, or more likely to laugh aloud. She loved jokes. Yet if something saddened her, some injury to a friend or the news of an accident, her face would fill suddenly with woe, and she might lay a hand over yours in sudden sympathy.

I never thought of Jenny as a sex object for myself. She was not a girl, but a woman, undoubtedly in her late twenties already, perhaps even thirty, the age of some of my schoolteachers, and at least ten years older than myself. The fact that her solid little body was gently curved about the breasts and hips, that her breasts were full and her mouth warm, affected me not at all. I liked her because she was so happy, so unreservedly friendly, and so full of wit.

One of my earliest experiences with Jenny shook me so it never left my mind. I had known her merely as one of the hearty ladies who worked in the laundry, who dressed in haphazard, shapeless clothes, and who appeared occasionally about the hotel to borrow a boat for an hour or two, or to wet themselves up to the knees in the water. Jenny and I had become friends when I had "rescued" her from a leaky boat after she had pushed off from the dock without oars. I had simply waded out, without regard for my trousers, and pulled her and her boat and her fat lady companion back to shore. And she had faithfully promised me that she would buy me a pair of socks for saving her life, a promise she never kept, although she usually slapped herself on the forehead and reminded herself of it whenever we met.

I had known Jenny no more than a week when she invited me one day to ride along with a big square-faced man and two other women from the laundry to a nearby stream where the smelts were running. There we watched as men with enormous wire baskets roped to the ends of long poles dipped up smelts by the gross and dumped them into boxes. After a time we each took turns wielding the baskets and heaving the thrashing messes of tiny fish up out of the chill water. It was a wet day and there was mud under our feet. But there was a small bottle of raw white liquor passed around to warm us, so before long we were merrily insulting each other and hailing each other by name as if we had been brought up in the same yard.

I walked with Jenny down the woods trail back to the car and while we stood there awaiting the others, I noted a small hole in a tree trunk, as perfectly round as if it had been made with a tool, yet grown over with bark all about the edges.

"What the hell is that?" I asked her. "Did a woodpecker do that?"

Jenny, her eyes bright, grinned up at me.

"Oh, that's just where those old lumberjacks go. One of them wooden wives."

I felt myself blush scarlet, ears, neck, and face right up into my hair. Jenny's expression at once grew tender. She laughed gently, took me in both arms, and squeezed me tight. Like many working women in that day, she wore no scent but her own body odor, not a thoroughly pleasant smell but not repellent either. I felt as if I were being hugged by my mother.

"Oh, Jesus Christ!" she murmured. "We're so *god*damn young!" Then she pushed me away and laughed right into my face. "I got no goddamn sense at all," she said.

Jenny, at this time, Jim told me, was staying in the Ryerson cabin, a tiny red cottage near the narrow outlet of Rangeley Lake, made up of two rooms and a shed, with a little hovel out back big enough for two horses. There was a small group of souls who all lived thereabouts, seeming to swap cabins about or share them from time to time, and all were related in a way I could never completely unriddle. The oldest of them all was Henry Baker, a teamster as bent and bony as a Halloween witch, who would answer a simple "How are you?" with a conscientious listing of his every ailment, beginning always with, "Wal, I ain't so good." Henry was married to Jenny's sister, who was at least forty years younger than he. Ernest Ryerson, the teamster who owned the red cottage, was an ex-farmer in his thirties, who bore a distant relationship somehow to Henry, perhaps through one of Henry's earlier marriages. And Jenny was married to Myron White, who drove the motor-bus "stage" between Rangeley and Oquossoc. Myron himself, to my recollection, was either the brother or brother-in-law of the pretty young woman who had been Ernest Ryerson's wife until she took up with a tall gloomy man who carried her off to California.

Henry Baker owned a big house on a side road not three hundred yards from Ryerson's where sometimes you might find the whole group gathered. All of them, like every family in Oquossoc except two or three (out of twenty-four), pieced out their earnings in the summer by trafficking one way or another in liquor, either transporting it, peddling it, or actually brewing it in earthenware crocks. But they all worked hard at other callings, winter and summer, about the hotels and lumber camps or the steam laundry.

Daylight hours in December fled so fast that if I took time to call on Jenny it might be black dark before I could make it back to camp, where I would surely have to give an accounting of the extra time it had taken me to perform a few simple errands. So I never would have seen Jenny at all had she not appeared at the railroad crossing just after I left Jim. She wore a short brown coat that had been made for a man and a hat like a bucket that hid her hair and eyes completely. Yet I knew her at once by her bouncing gait and she knew me too, for she hailed me from several rods away, waving one hand and tipping her head back so she could see under her hat brim. I waited in the wide road before the store until she came up to me, then took her by the hand. "Old man Smith!" she greeted me and offered me her enormous smile. Her eyes gleamed at me from deep under her hat. "Jesus God!" she exclaimed happily. "You going to keep them?"

I supposed she meant my whiskers — a scraggly growth that flourished only on the very end of my chin and thinly on my upper lip. Instinctively I rubbed my fingers over my face.

"Why not?" said I. But I really meant to get rid of them as soon as I could find the courage to shave.

"Oh, they're pretty!" said Jenny. "They're kind of pink." She still held tight to my hand. Her grip was warm and meaty as a man's.

"Jenny," I said, suddenly inspired, "you want to do me a hell of a favor? You going into the store?"

Jenny nodded, still smiling, still gripping my hand.

"This old bird up at the camp wanted me to buy him some goddamn talcum. I feel like a fairy to ask for it. Look." I pulled the talcum can partway out of my shirt and showed it to her. "How about you buying it?" Jenny reached out and took the can away from me, before I was ready to let it go. She held it up where the whole world could have seen it, had they been watching. "Ador-me," she murmured. "You mean one of them hairy old lumberjacks uses this? Not that old Paul Cyr? They's not enough in one can to make *him* smell sweet."

"Gus Lorry," I said, and promptly wished I had never spoken, for Jenny's face seemed to expand in almost joyous amazement.

"Gus *Lorry!*" she cried. "That old spider! *He* uses *this!* Well, Christ all Jesus! He must smell like a polecat without it!"

Jenny and I, in the manner of many country people, had been talking to each other as if we were still fifty yards apart. Our two voices rang over the rooftops of the store and the station and the little square barbershop that stood right across the road. I pulled my hand out of hers.

"Christ!" I whispered fiercely. "Don't tell *everybody*. I never meant to say his name!"

Jenny grew instantly contrite. Her wide soft mouth hung open and she set one hand over her lips. "Oh, I wouldn't ever say!" she assured me earnestly, in a quiet tone. But she had already *said*, loud enough to inform the entire resident clientele of the Dead Rat. I shook my head wretchedly and Jenny, looking as if she were ready to weep, put a mittened hand on my arm.

"Oh, nobody pays any mind to what *I* say," she murmured. "That's a fact."

Whether it was a fact or not, I was comforted and we walked together into Mackenzie's. It was warm in here, where an oversize heater stove always burned steadily, no matter the weather. There was just one lamp lit, back in the corner in the section that had been walled off for the post office. An unseen hand behind the partition there kept popping random trophies into the tiny glass-doored letter boxes that opened out into the store. Cliff Hill, wearing a wool hat against the drafts that haunted the barn-high room the store was made of, stood back in the shadows behind the counter, close to the cash register. A tall man with a body massive as a bear's, wearing a shiny leather jacket and a rolled-up stocking cap, faced Cliff across the counter, and both turned to eye us as we came in.

"Well, Jenny!" Cliff called out. "You keeping cool?"

"Oh, I stay out of the sun," said Jenny, with affected solemnity. "Nighttimes, anyway."

"Well, hell," said Cliff, "whyn't I just run down there and warm you up tonight? Might be able to get away. I could bring Eben along to spell me. Right, Eben?"

The big man, whose large lips shone wet as if he had been sucking a lollipop, exposed his small brown teeth in a grin.

"Do my damndest," he said.

Jenny kept her eyes on the grocery shelves, as if she were seeking the right label.

"I got bear traps set in the dooryard for critters like you," she replied, and both men laughed more loudly than the remark deserved. Unacknowledged by either man, I had gone straight on down to the post-office wicket to wait for it to be opened, fishing meanwhile from my jacket pocket all the letters I carried. Down here in the semidarkness, by the meat block and the cookie boxes, another man was lounging, apparently waiting for the mail to be distributed. He told me howdy and I murmured a polite reply. I had often seen him in Oquossoc in the summer and knew who he was but had never had any conversation with him. He was Johnny Foster, head clerk for International Paper, a man in his late twenties, very dark and lean, with a bright sharp face led by a nose of more than average length. He was good-looking in the swarthy slick-haired style that was fashionable then. He had lost an arm in a woods accident and carried one mackinaw sleeve now tucked neatly into a side pocket. He wore a limp felt hat set back on his head, revealing two wedges of polished black hair. Handsome, self-possessed, and accomplished (men had often in my hearing remarked on Johnny's quick climb to a sort of executive job with the I.P.), Johnny seemed to me then exactly the sort of youth to win the heart of Cliff Hill's daughter. I stood back from the wicket, to allow him first access to it when it opened, and I glanced at him glumly now and then, while I pretended to be studying the letters I held in my hand.

"You come down from McCormick's?" Johnny inquired cheerfully. I nodded.

"How do you like?"

This usage of "like," without the "it," employed to mean satisfaction with a job, was one that had always mildly grated on me. I shrugged.

"Oh," I said. "All right." I gloomily treasured this symptom of Johnny's backwoodsiness, as a point that might cost him favor. But Johnny was determined to be friendly.

"Pair of bastards, ain't they?"

Privately, I agreed. But I could not still a conviction that it was traitorous to blackguard the boss in front of a rival. I.P. was the *other* paper company, whose pulpwood might become mixed with ours. I shrugged again.

"They're kind of tough to work for," I granted.

"Bet your ass," said Johnny. "You hear about the shooting?"

"Shooting?" I shook my head. My heart actually leaped at the thought that the McCormicks might have been shot at. Johnny lowered his voice.

"Eben Kendall's down there gassing about it to Cliff. Somebody shot Jack Dunham right through the heart. Deader'n hell."

I knew Jack Dunham only by name. He and his wife operated a store at Haines Landing.

"They know who did it?" I asked, keeping my own voice low simply because Johnny had.

"Couple of Frenchmen, according to Eben. I don't know how the hell he knows."

The outer door opened then and admitted another man, obviously come for the mail, for he headed straight down toward us.

"Hawkshaw in person," Johnny muttered.

Both Cliff and Eben greeted the new man heartily.

"Hey, Stape!" "*Mister* Stapleton!"

This was H. O. Stapleton, the local deputy sheriff, a very tall man in a wide-brimmed hat and a gray wool jacket worn unbuttoned. He had the weathered face of a lumberman, made a little lopsided by a small quid of tobacco he carried inside his cheek. He had extraordinarily clear blue eyes that he would widen as a sort of greeting when he met anyone. He nodded repeatedly to the two men and to Jenny, who had turned to face him, but he kept on coming straight toward me. And to my dismay, it became clear that it was me he had come to talk to. I had heard enough about Stapleton's cold-bloodedness, his willingness to take his best friend in if he caught him selling liquor, or to shoot a man who tried to escape him, so that I felt a trickle of fear as he came to face me. But Stapleton's greeting was mild enough and I was able to smile at him.

"You're clerking up to McCormick's this season," he informed me and I confessed I was.

"What've you got for Frenchmen up there?"

I was alarmed to discover that I could not count a single

Frenchman. How many? There was Paul Cyr. Who else? My face grew hot.

"I don't know for sure," I gasped. "We have a lot of Dutchmen." Of course I realized that Dutchmen wouldn't do and I flushed more deeply still. Thank God for the dim light.

"Well, Jesus! You got some!" Stapleton declared, smiling one-sidedly at my confusion.

"Well, there's Paul Cyr," I said. "The bucksaw man. And Robbie. Robichaud. A couple of others."

Eben Kendall had come up by this time to join us. He had a voice that seemed to rumble up out of his cavernous belly.

"I know that Cyr," he declared. "Great big feller. Twarn't him. That I know. *Couldn't* have been. These was medium-sized critters, built kind of hedgehoggy, like so many of them bastards. One had a full beard."

Stapleton kept his eyes on me, to see if this intelligence would help me name the men. When I did not respond, he prompted me.

"Any of them Frenchmen got beards?"

"Only Paul Cyr. He and the feeder are the only ones in camp that have full beards."

"What feeder?"

"Gus Lorry. But he's no Frenchman."

"Hell he ain't!" Eben Kendall bellowed. "Lorry! Lorry! I God, I'll bet you that was La-ray or La-roo, or some such a thing to begin with. Tell *me* he's no Frenchman! What's he look like, young feller?"

I could feel the sweat forming on my face. I looked desperately at Stapleton. Was I going to have to betray Gus once more?

"Christ!" said Johnny Foster. "I know that Lorry. He's no more a Frenchman than you and me. Maybe back a

86

hundred years or so. But there's a slew of Lorrys over around Errol and they sure as hell ain't French."

"He doesn't look a damn bit French," I added, trying to sound as self-assured as Johnny. "He talks English."

Stapleton laughed loudly at this, and looked almost fondly into my face.

"You sure he don't talk French too?"

I squirmed and shook my head. Eben Kendall had crowded close to me as if he were trying to force me off a plank. He was several inches taller than I and probably seventy-five pounds heavier. When he stood close, I could hear his breath whistling in his nostrils. There was a rank smell of old cigars about him, which drowned a faint odor of shaving lotion. I refused to back off for Eben, and my elbow, when I shifted by arm, banged into Eben's belly. It felt as if armor plate lay under his heavy shirt and mackinaw. He was a top-heavy man, with a chest that seemed to swell like a grouse's, and a flat belly. I realized now that he wore a sort of corset under his clothes.

"This Lorry," he demanded of me. "He about your build? Skinny feller? What?"

I gave my reply to Stapleton.

"He's shorter than me. Bony guy."

"Living skeleton," said Johnny Foster.

Now Kendall bent his baleful gaze on Foster. His eyes were dark brown. Behind the thick glasses the pupils seemed large and round as a dog's.

"By Christ, I'd just like a look at that feller!" he rumbled. "I'd know the son of a bitch if I seed him."

"He never comes to town," I said. "Not since he came to camp."

"You with him all the time?" Stapleton inquired mildly.

"Well, no. But I see him every day."

Kendall snorted and backed away. "See him every day,"

he muttered. Then he raised his voice and glared at me. "You see him every minute?"

"He can't leave the horses," I said.

"Oh, shit!" said Eben. But Stapleton nodded, apparently satisfied.

"I'd like to have a look at this bastard!" Eben growled.

"I'll take a look at him," said Stapleton. "Won't be hard to find out what he's been up to."

My heart sank at the thought of my leading Stapleton into camp to meet Gus, as if I had betrayed him to the law. But Stapleton had far more patience than Kendall.

"That team'll be coming down from the storehouse tomorrow. I'll take a ride back with him when he goes. Come by about noontime, don't he, sister?" Stapleton addressed this question to Sally Hill, who had quietly slid open the wicket and watched us gravely.

"Around there sometime," she said, in her little-girl voice, looking right at me as she spoke.

"Well, I'll make it my business to be right here when he comes," said Stapleton. He looked intently at Kendall. "Don't fret. I'll have a talk with that feller."

"If he's still there," Kendall muttered.

"You want I should go up and take him now?" Stapleton said testily. "You ready to swear out a warrant?"

Eben pulled his mouth down unhappily.

"Well, I'll carry you up myself in the morning, if it comes right so's I can get away. I don't even know what the bastard looks like."

"Nothing hedgehoggy about *him*," said Johnny. "Looks more like a shit poke."

"Johnny *Foster!*" Sally gasped. "Don't be so *awful!*"

Johnny pretended to dodge away, as if Sally had aimed a blow at him. His glance lit on Jenny, who was grinning her wide grin. "That's right, ain't it, Jenny? You know Gus

Lorry. What's he look like to you? You ever see a man looked any more like a sh . . ."

Sally silenced him with a scream. "Johnny!"

Everyone laughed now except Eben, whose round, coffee-colored face had begun to shine with sweat.

"Well, this feller was more on the stout side," he grumbled. "Bushy beard. Kind of hunched-up in the shoulders."

"Well, that's not Gus Lorry," I declared. Jenny and Johnny made noises of agreement.

"Seems as though you need to make up your own mind, Eben," said Stapleton, "before you set me to dragging in suspects."

"*I* didn't ask you to drag him in!" Eben cried, in a voice made high-pitched by injured innocence. "I never set *eyes* on the feller. Don't sound from what these fellers say he looks anything *like* the ones I seen. Alls I said was these was *Frenchmen*. They *looked* French. They *talked* French. That's all *I* said."

Stapleton started to move away now, looking off through the window and fishing absently in his back pocket.

"Well, we'll get you a Frenchman then. Too many of them bastards in the country anyway."

Eben was still mumbling, like a kettle that had been just lifted off the stove. Johnny accepted the bundle of letters that Sally handed him through the wicket and I moved in as he walked away. Sally's eyes met mine. Even now, when I cannot always summon up a picture of her face, I can recall the tingle that would reach down through my insides, like a searching finger, when I looked directly into her eyes. It was not that she was the prettiest girl I had known, although she was pretty enough. Her looks offended me sometimes, for her cheeks were often soiled with rouge, which she was too young to have any need of, and sometimes she tried to draw a tiny red mouth with lipstick on the center portion of the generous lips God gave her. But when I looked straight into her eyes and discovered all over again how very deep and clear they were, it was as if we had each told the other some soul-deep secret we dared not say aloud. The effect on me was to rob me of words entirely. I dared not believe that Sally was moved at all, for I had often heard her laughing intimately with another young man and had seen her lay hold of some fellow's arm and clutch it as if he was dear to her. Nevertheless I cherished her attention and conspired to lengthen the time we stood close.

"You have a package to sign for," she told me, in the baby soprano many girls in that day felt impelled to use when talking to any man they were trying to attract. She took a small box up from somewhere and held it on the far side of the counter while she pushed a white slip toward me, along with a stub of pencil. I took the pencil and she thrust her hand out against mine to direct me to the proper line. "Right there," she half whispered. And even while I winced inwardly at the way she said it — almost like "wight dare" — I held my hand against hers an extra second, as if I needed guidance to the spot for my signature. Then I signed it, and she took the pencil and slip back from me in such a way that our fingers were entwined for a moment. I accepted this opportunity to squeeze her small sturdy fingers ever so lightly.

"You're *fresh*," she whispered. But she did not immediately pull her hand away. And glowing with that small conquest I took the package somewhat blindly and started off.

"Don't forget your letters," she sang after me. I immediately flushed scarlet and came back to gather up the assorted mail clumsily in one hand, only to drop part of it to the floor and have to set the package down to bundle it all up again. Jenny was right there to pick up two of the envelopes that had skittered away and she laughed gently as she gave them into my hand.

I plunged out the door then, hardly noticing the other folk who had gathered in honor of the fact that the mail was "up." Outside in the wide snowy square I waited for Jenny, who came carrying a large brown paper bag full of groceries, from which she drew the can of talcum powder and handed it to me. I tucked it quickly out of sight inside my leather coat and shirt.

"If anybody should ever ask me, which they won't," said Jenny, "I'd say that old bullfrog of a Kendall killed Jack Dunham."

"Kendall?" The thought startled and enthralled me. He had always seemed a repulsive man to me — not just from his being an undertaker, but because of his general arrogance toward all lesser folk and the fact that his mouth always seemed to be greasy from his last meal. I habitually pictured him pushing through doors regardless of anyone who might seek to precede him, reaching out over the dinner table to draw all the side dishes within his own reach, and hacking open corpses to despoil their intestines with his soiled and puffy hands. But shooting a man deliberately through the heart?

"What would he kill Jack Dunham for?" I whispered. I could not see why *anyone* would kill Jack Dunham, a mild, smiling, round-shouldered man who always seemed to go about on tiptoe.

"Who the hell do you suppose does up Kitty Dunham's chores while Jack's off to these sportsmen's shows?" Jenny muttered earnestly into my face. "Why that old son of a bitch hardly waits until the door shuts behind Jack's britches when he's up there putting the boots to Kitty. Him and that other old caterpillar, that Harland Bailey. I've seen the two of them sneaking up them stairs behind the store, one after the other like a couple of overgrown kids, bringing a goddamn bottle. I suppose they took turns in bed with her. You imagine being in bed with that Kendall? That Christly hoptoad?"

"What do you suppose? He caught them at it?"

Without having meant to, I was plodding at Jenny's side down the rutted road toward the Ryerson cottage.

"Oh, hell!" said Jenny. "He must have knowed for years what's been going on. Why they'd be upstairs making that goddamn bed creak like a buckboard and Jack'd be trying to wait on trade down below, white as a junk of suet. Tears in his eyes. He *had* to know."

"But Jesus. What would they kill him for?" It was impossible for me to picture Kitty Dunham as a woman anyone would fight about, much less commit murder. A hard-bitten, narrow-eyed woman who looked about as sexy as a bolt of firewood.

"Oh I suppose she wanted to be shed of him and she talked them into it. Unless maybe she held the gun herself. I wouldn't put it past her. I seen her whip that dog of hers until the poor critter could barely crawl."

I realized that, with the sun getting low, I was moving too far in the wrong direction, so I stopped in my tracks. Yet my thirst for details of this business, even though they might all be Jenny's imaginings, was not half slaked.

"How the hell do you know all these things?" I asked her. "You're not making them up?"

"Everybody knows about it, everybody at Haines Landing anyway, that has any eyes in their head."

"How come Stapleton can't figure that out? He's looking for two Frenchmen."

"Oh, Stape ain't been out of the woods long enough to get his bearings. He's still looking for tracks in the snow. But I tell you one thing. If he gets anything on Kendall, he's not going to let go. He's the only man in town ain't half scared of that big shitass."

"Jenny," I told her, by way of farewell, as I started back toward the village, "you don't talk fit to eat."

Jenny aimed a feeble slap at me.

"That's from listening to critters like you. But you see if I ain't right about Kendall. Not that it'll ever come out. They've had time to cover up all the tracks."

Remembering Kendall in the store — arrogant, blundering, childish, yet simple too, and no different from half a dozen everyday men I had dealt with, no more menacing than Tim McCormick — I found I could not think of him as a murderer at all.

"Oh, I think you've been hitting the pipe, Jenny," I called to her, laughing.

Jenny was never offended. She stood ankle-deep in a patch of snow and grinned at me.

"Well *somebody* put a bullet in him. Jack Dunham never shot himself."

Well, that was a fact. I began to fit the letters and papers into the pockets of my jacket and tucked the small package — it was addressed to Mrs. McCormick — under my arm. Then I started on the long hike toward camp.

"You stay clear of that little postmistress!" Jenny cried. I wriggled inwardly with delight to be linked with Sally this way. I waved at Jenny and left her standing still behind me, an odd little figure, sturdy, all misshapen in her outsize clothes, and the bucket of a hat that drowned her eyes, yet winsome too. I strode quickly along the road, treasuring the tale I would have to tell at camp that night.

Dunham's little store, which was busy only when the vacation crowds were here, stood but a few rods from where the winter road by Indian Rock left the main highway. I could actually see the small building, faced with creosoted logs, with a long green-roofed porch the length of the front, when I reached my turnoff. Without pausing for an extra thought about it, I kept going past my turn to look at the building where the murder had taken place. It was still and empty, the front steps unshoveled, the walk deep in undinted snow. But someone had made a narrow path to the outside stairs that led to the living quarters over the store.

I had no idea what to look for, nor have I any recollection of what I hoped to see. From the chimney out of sight behind the steep asphalted roof, a thin gray smoke twisted skyward. I merely stood there and stared, perhaps wondering if a ghost would materialize behind the black window to mouth his dark tale of betrayal. And as I stood there, I suddenly became aware of a movement behind the glass upstairs and I looked up to meet the stare of Mrs. Dunham, her hard face grimmer than I had ever seen it, her eyes squinting venomously directly into mine. I made a foolish attempt to pretend I was looking all about for a man who did not exist. I turned completely around and peered into the scanty woods across the road, even took a step in that direction and craned my neck this way and that. Then I put on a broad show of disappointment, shaking my head, frowning, screwing up my face in disgust. At that moment I caught sight of a man coming out of the outdoor toilet. He stopped to look at me but I just turned back up the road, as if abandoning a search. A heavy car came lumbering toward me, its tire chains rattling in rhythm, wet snow spitting from its wheels. I stood on the roadside to give it room and it pulled to a stop just beyond me, right in front of Dunham's store. As it came opposite me a big coffee-complexioned face had turned an angry glare toward me, but the car was gone before I realized that the face was Eben Kendall's. It had actually been distorted with hatred or rage.

What the hell, I asked silently, have I ever done to you? Am I not even supposed to be in the road when you're driving? I looked back once and saw he had got out of the car and was standing there staring after me, his glasses gleaming blankly. He was a monstrous-looking son of a bitch, really, made twice as big by his bulky clothes and oversize shoepacs. For that second or two I was convinced he could have killed a man without a moment's remorse. Conscious of his eyes on my back I strode along earnestly to my turn and did not even glance back when I swung into the woods, lest I tempt him to look at me.

The sun was low, making long shadows in the woods, where the hardwood trees were already bare and only the dark evergreens still treasured some snow in their bowels. The birds were still. Only my own tracks had marked the snow on the tote road. This, I told myself, trying to laugh at the idea, would be a hell of a place to walk into that pair of bastards. But I decided there was nothing to laugh about in the notion. Suppose somehow they had convinced themselves that I was trying to play detective. Or that I had heard or seen something . . . Oh, shit, I told myself. That's just Jenny's fairy tales that have got me going.

But as I moved deeper into the still forest and the turn of the tote road left the highway out of my vision altogether, I grew more and more uneasy. When the gloomy array of cabins at the fishing club came into sight I took care to pick a path far to the other edge of the tote road, practically in the brush, where I would at least have warning of an ambush. (Kendall *could*, I reasoned wildly, have cut through the woods, to overtake me.) And when I saw, across the road, in the brush, the deep fresh marks of a man's feet in the snow, real fear took hold of my heart and squeezed it almost dry. It was nearly a minute before I realized I had made the footprints myself a few hours ago when I had walked off the road to take a piss.

Oh, Jesus! I said aloud and laughed at myself. But the fear did not all drain out. I still watched the squat buildings as if they might spring to life. And I hurried on past at a pace just short of a trot, making the corduroy bridge rattle.

When I came to the spot where my tracks took the trail to the lakeshore, I stopped and looked about. Of course the idea was crazy — that Kendall and Mrs. Dunham might have come up over the ice to waylay me in this stretch of dark woods. But the sun was behind the mountain now and it was dusk where the growth was thick.

It was immensely still, with the winter stillness that seems to hush everything alive. When something did move at last and a tiny bird flitted close to my head, to look right into my face and inquire: "Dee-dee-dee?" I was as startled as if a human face had appeared. I could hear my own heart knocking gently at my ribs and echoing in my eardrums. I plugged along then, eager to get out on the vast white reach of the frozen lake, where it would be daylight still, and where I could see everything that moved within a mile. I could not keep myself from looking back when I had opened about a half-mile distance between myself and the land. The alders were thick along the shore behind me, making a long gray streak at the edge of the ice. And there was something there! Had it just appeared, rising dimly there where the woods began? I did not remember seeing it when I glanced back once before. I turned full around and stared for about three seconds. Could it be just a stump, darkened somewhat by melting snow? A wayward deer? It did not move. Or did it? I dared not watch too long. And when I started off again I was seized with the crazy notion that I should perhaps duck low and take a zigzag path. Oh, for

Christ's sake! I told myself. What the hell would they *shoot* you for? So I kept going straight for camp, at my steady pace. After I had traveled perhaps ten more rods, I ventured another backward glance. This time I froze, for what I had seen before had vanished altogether. So it must have been alive!

Jesus! I felt my chest grow tight. I stood still and studied the shoreline all along. There was no movement anywhere. A mile below where I had left the shore, a brown cabin stood, its windows boarded and its roof jack aslant. I kept watching to see if any figure would show up in the woods above the shore, on the cabin porch, or atop the distant ice. But nothing appeared, although I watched for two or three long minutes.

Trudging then over the solid ice, I decided finally that it *must* have been Kendall I had caught sight of. He would have followed me out, after talking to Mrs. Dunham, to make sure where I was headed — that I was not just ducking off in the woods, to sneak back and pry some more. And I told myself earnestly I had better not talk about the shooting any further, not even in the privacy of the camp, for now men might walk out to town and back whenever they had the time and the will and I wanted no gossiping in Oquossoc over any suspicions I might harbor. Certainly I must not breathe what Jenny had told me.

Yet when I found myself back in the office camp at dark, with the grateful heat throbbing in the stovepipe, I nearly burst my chest trying to stifle all I had to tell about. The supper call came soon. There was seldom much talk at meals save for loud requests to pass this or that, or to fill a pannikin. Most of the noise arose from the rattle of forks and spoons on the tin plates and the banging of potlids, and the shifting of feet.

After supper, there was often a gathering of some sort in

the office camp. Tim invariably put his glasses on and settled on his hard chair to read. Wallace would appear with his lantern. And one or two others might come in to buy tobacco or a whetstone and sit for a few minutes on a bunk to make a show of fellowship, although there were not often more than a dozen words exchanged.

This time, Harold and Tim and I sat there with no one else for company, Tim reading and Harold reciting some story about a dance he had gone to in Errol, months before. Tim made an occasional pretense of listening, putting his magazine aside and peering lugubriously over his glasses to offer a feeble laugh or grunt. The story soon petered out and the need to tell something of what I had learned in town at last overcame me.

"There was a shooting in Oquossoc last week," I said, trying to sound entirely casual, as if I had only just recalled something I had heard. But Harold sat suddenly erect and Tim laid down his magazine and took his glasses off.

"Shooting?" Tim said. "Somebody hurt?"

"Jack Dunham. He was killed."

"Dunham," Tim repeated. "That joker run the little store at Haines Landing?"

"That's him," said I. "They said he was shot right through the heart."

"Who the hell shot him?" Harold said. "Some crazy hunter?"

"They said somebody did it on purpose. They're looking for a couple of Frenchmen."

"Well, Jesus Christ," said Harold. "Why would anyone shoot the poor bastard? Were they trying to rob him?"

I shrugged.

"Nobody knows. Well, I guess they figure it was a robbery. I didn't get the whole story."

Tim shook his head to indicate bewilderment or sorrow.

"Too bad. Too bad. But they couldn't have got a hell of a lot. He must just about starve to death in that store."

Before I was tempted to tell more than I meant to, Wallace came in with his eternal gas lantern blowing. Tim told him the news before he had got the door closed behind him.

"Where'd you hear that?" Wallace demanded and Tim indicated that I was the source. Wallace pursed his lips tight and looked at me as if he might decide not to believe any of it.

"Shooting? In *Oquossoc?*" He seemed to be implying that there was not gumption enough in the whole town to get a gun loaded and fired. "Be damned. When did this happen?"

"Oh, last week. Johnny Foster told me."

Wallace made his mock-innocent face.

"Oh, Johnny *Foster*. Well, if *he* told you, I'd have half a mind to believe it. Johnny *Foster* wouldn't say anything that wasn't the God's honest truth."

"Oh, it was true," said I. "Stapleton was asking around the store about anybody seeing some Frenchmen."

"A Frenchman shot Jack Dunham?"

"That's what they think."

Wallace shook his head solemnly and took his accustomed seat, with the gas lantern placed tenderly between his feet.

"Jack Dunham," he murmured. "Well, that ain't much of a loss."

Harold and I exchanged a long look at this; but no more was said. Tim did not resume his reading. Instead he stared meditatively at the stove, twisting his head and grimacing slightly, obviously feeling it was his duty to offer more than these few sentences to do honor to the dead.

While we all sat silent, Gus Lorry burst in, his eyes as

wild as ever, his face wearing its usual expression of frantic dismay. Yet he had no message of doom at all. He panted a greeting to everyone, then held his soiled hands over the stove and rubbed them together.

"Fuckin hovel is *cold!*" he declared, and no one contradicted him. Tim picked up his magazine and seemed ready to burrow right into the page. But Wallace watched Gus with a faint expression of distaste.

"Something we can get for you in the wangan?" he inquired sharply.

Gus widened his eyes even more and shook his head.

"Noooooo. Noooooo," he said. "I just stopped by." He studied the stove top as if he had something cooking there. His hands kept caressing each other. "Less'n you happen to have another plug of that black Planet."

Wallace glanced at me but I was already on my feet. I was the only one, besides Gus, who knew what he really had come for.

"I can probably make you a deal," I said. I led him into the tiny room at the back of the cabin where the wangan goods were stored. As I made a show of digging about under the counter to find the carton of chewing tobacco, I slipped the talcum can from under my shirt and set it on the counter, well out of sight from the other room. Gus deftly retrieved it and snuggled it somewhere under his reeking jacket. Then I handed him a ten-cent plug of the ink-black tobacco and he fished deep in his pockets to bring up one nickel at a time. He moved back into the front of the cabin, absently nibbling a tiny chew off one corner of the plug, then stood again by the stove, rather ostentatiously working his cud around in his mouth, to make sure, I suppose, that no one would imagine he had come here for any reason but to get a mouthful to chew. And I, like all amateur conspirators, set out at once to urge attention away from what Gus and I had been up to.

"Saw an old friend of yours in town today," I offered, then wished to God I had sorted out some other sentence before opening my mouth. Wallace wet his lips, swallowed, and seemed to be forming words silently, which he kept pressing back into his mouth. Gus regarded me with joy, as he did anyone who ever invited him into conversation.

"Who's that?"

"Jenny White," I told him, already flushing slightly as I contemplated the embarrassing turns this gambit might involve.

"Jenny White!" Gus cried, with such force that even Tim was startled into lifting his eyes from his magazine. "Jesus! You know Jenny White?" Gus turned to each man in turn to invite their separate replies. "You all know Jenny?"

"Not me," said Harold. Wallace merely shook his head, not I am sure to deny he knew her but simply to encourage a prompt end to the discussion.

"I hear she's pretty hot stuff," said Tim, who never could resist pursuing any discussion of sex, no matter what the source.

"Hot stuff!" Gus yelped. "Hot stuff? Well, you can just bet your hairy old ass she's hot stuff! Why when I . . ." Gus momentarily choked as the words crowded to his lips. ". . . when I . . . Jesus Christ Almighty! Why she comes off three times to my one! By God and by Jesus! Three times to my one!"

As he finished that declaration, his eyes met mine. And instantly I knew that he was lying. And I knew that he knew I knew it. He subsided then, but kept muttering: "*Three* times, for Christ's sake." Then he told us a sudden good night and dodged out the door.

❦ 4 ❦

THE first snow fell before Christmas, beginning late at night, after all the lamps were out. The wind hardly blew and the snow fell without sound, so I was startled, on opening the cabin door in the predawn twilight, when a small bolster of snow dropped in and shattered silently at my feet. The cabin steps had become a high banking of snow, sloping without a break to a ghostly desert that, in the uncertain light, seemed to reach as far as a man could see, making one enormous world of the camp yard and the woods and the lake and the hills, where only myself was stirring.

Anyone who stepped out there to find the outhouse would likely sink in snow nearly to his neck. I reached for the stubby broom and managed to shove a sort of path through the drift, sufficient to let me out where I could at least relieve myself without soiling the steps. The day had not really begun, yet the sky, all clouded still, had grown faintly gray.

Without a word, Harold, in unlaced boots, came out to join me. We stood there, two skinny figures in our union suits, and reverently pissed two divergent dark streaks in the new snow. There seemed to be a faint stirring in the cookhouse, yet no sliver of light showed. It was several seconds before we realized that the uphill windows of the cookhouse and barroom had vanished as if they had been sunk in the lake. The snow reached in gently rolling dunes clear down to the ice, burying buildings and brush, with only the warmly breathing roof jack poking up like the mast of a sunken ship to indicate where the kitchen lay.

"Christ all Jesus!" Harold whispered. And right then the cry of the cookee sounded, faint as a call from a well, announcing the day to the sleeping barroom: "Turn aaaaaa-ooooot!"

After a few seconds, as a faint rumbling began in the barroom, we could hear the cookee fighting his way out the shed door, perhaps kicking the hip-deep snow aside. A yellow glow that seemed to have no earthly source faintly altered the surface of the snow about where the shed might have been. The voice of the cookee burst out more loudly now, clear as an icicle, yet strangely softened, as if each syllable had been wrapped in wool:

"Turn AAAAAAAA-OOOOOOT!"

Inside the cabin Tim struck a match and we looked in to see his red face vignetted in the tiny light, his gray hair tousled and his eyelids still puffy from sleep.

"What about closing the fuckin door?" he croaked. He was trying to get the lamp wick lit and the draft from the door kept twisting the little flame awry. Harold and I came back inside and pushed the door to. Harold packed paper and kindling into the stove and set a fire going. We both stood close and tried to believe we could feel some warmth as the new flame felt its way among the splintered kindling. Tim cleared his throat and began to cough, as he did every morning before he had even got out of his bunk. "Son of a bitch!" he told himself in a strangled voice. He rose wearily to his feet, still wearing his socks and wool shirt, and padded over to spit in the stove.

"Snowed more'n three feet!" Harold announced. Tim blinked owlishly at him, then moved to the door, bowlegged as an ape, his soiled long underwear all out of shape at the knees and lodged partway up his shins. He stared out into the dark, then beat his fist against the doorjamb.

"Son of a bitch to Jesus!" he growled. "They won't cut another fuckin stick!"

Harold and I looked at each other as we so often did when we shared some thought in Tim's presence. We both knew there were damn few sticks left to cut, beyond some eight-inch spruce high on the ledges and one fir thicket of little account. Tim could have sent all the choppers home tomorrow and left the rest of the cutting to Paul Cyr, who was just wild enough to wallow out in snow up to his chest and clear a space to drop all that timber down and limb it out for his bucksaw. Even the boom logs had long since been limbed out and scaled. All the Dutchmen had been making ready to leave, except for Harry Ernst and another one or two who would stay on as teamsters. Paul and the two young Frenchmen who had come in to clean up the scantiest cuttings could handle all that was left, provided the snow melted down enough, as it generally did at this season. The teamsters would be on hand soon after Christmas and the McCormicks could pick up hands enough locally to man the main road when it was broken out, to scatter hay to hold the sleds back on the grade, and another few to work the snub lines where the slope was steepest.

But Tim kept grumbling as he drew on his britches and dressed his feet. The cabin warmed quickly. Harold and I pulled our long wool socks nearly to our knees and each put on an extra wool shirt to wear beneath our jackets. All in a file, with Harold going first, we wallowed into the snow, laboring like winded horses the thirty or forty yards to the door of the cookshack. It was bright and hot here, with all the big lamps lit and the cookstove already aglow. The air was thick with the smell of bacon and wood smoke and soiled wool. The long tables were crowded, the men, mostly with hair uncombed, all bending to their plates and pannikins without taking any note of our entry.

After breakfast we learned what might have been one reason Tim had cursed the snow so bitterly. He had forgotten to have the bags of soft coal hauled from where they had been dumped near the wharf up to the little bin by the blacksmith's forge. This was a job he had taken on himself as being the only man in camp with sense enough to see it was done. Now the bags all lay buried out of sight, no doubt frozen to the icy ground, so they would have to be dislodged with a cant dog, if indeed they could be found at all. Tim, wrapped to his eyes in mackinaw and stocking cap and dusted right to his shoulders with snow, poked angrily about with a long stick, kicking a pathway for himself and staggering often on the uneven footing. The curses, muffled by the collar of his coat, rose up as steadily as the smoke of his breath. While he still plunged about in the drifts, finding nothing but buried boulders under his probing stick, Wallace appeared, wading along on an ancient pair of snowshoes, with rawhide clips that had been broken and knotted again a dozen times.

"The coal?" said Wallace. Tim had no patience to reply and Wallace hastened to mollify him. "Seems as though Bob could have kept that in mind! He don't have too much else to busy himself these days!"

Harold and I, standing in the shed entry to pass some time with Andy, the cookee, had been watching Tim's every move. But Wallace had not seen me there. *Me* keep it in mind? For Christ's sake! My astonishment and anger showed so plainly on my face that Andy ducked his head and grimaced, as if making ready for me to explode. But I was too young then, as I am too old now, to blow my stack too quickly at my betters. Instead I simply looked at both Andy and Harold in angry astonishment and called them to witness that Tim had himself vowed that *he* would see that this job was done.

Wallace noticed then that we were standing there, almost hidden by a high drift, and he slogged over to us, crunching the snow under his webbed shoes.

"Couldn't you give him a hand there?" he demanded of me. "I should think you'd have had that coal *up* there before the *snow* come!"

"Tim said *he'd* take care of it," I told him sulkily.

"*Tim* said! *Tim* said!" Wallace echoed. But he seemed unable to discover a bitter enough rejoinder. He was certainly not going to belittle Tim's authority. "Well, you was there when they unloaded that coal," he went on finally. "You must have some idea where they dumped it."

I started to protest that I had not been anywhere at hand when the coal was unloaded, for it was brought up in the middle of an afternoon in November, when I was in the woods. But it was a fact that I had seen it lying there by the wharf for weeks, just as everyone else in camp had. But to find it now, with the landscape all distorted by the snow until the lake could not be separated from the shore! I took a broken cant dog from the shed and floundered out toward where Tim was angrily jabbing the snow with his stick. Harold came after me, for he had no more work to do, now that what little wood still needed scaling had been buried deep out of reach of his rule. Harold brought along a mop handle and began stabbing recklessly here and there to locate the prize. A cant dog was really a hell of a thing to try to probe with, for it is not particularly long and is made to be taken in two hands, but it had been handiest to me and I tried to make it do, lunging sometimes almost on my face when the cant dog failed to touch bottom.

Harold and I made a game of it, shoving each other aside to be first to probe a new drift and laughing when one of us lost his balance and fell sideways in the snow.

Wallace, who had undertaken to help Tim by taking the pole from him and probing more systematically foot by foot along what he took to be the shoreline, stopped his work to watch us with intense irritation.

"If you'd paid some attention to *business*," he snapped, "we wouldn't have to all be out here wasting valuable time. And if you'd been born with a little *brains* you'd have something besides a goddamn cant dog. . . ."

For some motive I never had the patience to delve into, any suggestion that I lacked good sense was the one insult best calculated to unseat my reason. I felt the hot blood surge into my face and neck and set my ears on fire. Sudden angry phrases so glutted my throat and mouth that they burst out like a sob. I lifted the cant dog like a spear and hurled it with all my strength off over the endless snow-laden surface of the lake. I did not realize then that work on the hayfields when I was a boy and a job rowing two hundred pounds of garbage in a heavy dory to dump far out in the bay had given me unusual strength in arms and shoulders. So I was actually aghast when the cant dog, instead of flying three or four yards off, went sailing on and on like a nightmare arrow to fall far out on the lake, without a sound, and with just a tiny puff of snow to mark its landing. It must have traveled thirty yards in the air. I looked open-mouthed at the other men and they looked open-mouthed at me, all of us too startled to speak.

"Jesus CHRIST!" Harold breathed, at last. But Tim and

Wallace seemed struck silent with fear. I was afraid myself, for my anger had dissolved and I knew not what I might have let myself in for. But I felt I had to complete my gesture and all I knew to do was turn and plunge off, trying to find some sort of pathway through the trodden snow. I had gone about halfway back to the cabin, where I suppose I would merely have sulked on my bunk for a half hour or so, when Harold called out: "Here it is! I got it!"

It seemed perfectly manly then to turn and go back to share the small rejoicing. Wallace fetched a shovel and quickly dug enough snow away to expose the tops of the three burlap sacks, all snuggled together in a frozen embrace. Without a word spoken, I joined in the effort, using a piece of plank I took from someone's hand, to pry the sacks free of the ground. Finally I had to trudge out on the ice, through thigh-deep snow, to retrieve my broken cant dog. Tim actually grinned at me when I carried it back. And Wallace made a small joke about being thankful it had not lodged high in a tree on the other shore. Harold and I, prying together on the cant dog, broke the sacks loose. Then Wallace fetched an extra pair of snowshoes, which Harold fastened on his feet, and together they began to tread down a pathway to the forge. They made five or six back and forth trips before Wallace was satisfied they could haul the sled without slumping. No one asked me to give a hand with the sled, but I took hold anyway, hitching the rope behind my neck, then around my shoulders in front and under my arms, while Wallace pushed the sled with a stick from behind. I sank in snow to my knees four times on the first trip but then the snow held up. We could haul only one bag at a time but we had it all safely in the bin within twenty minutes, then all of us, Harold, Wallace, Tim, and I, sat almost snuggled together to drink tea in the cookhouse, with our jackets undone and snow melting from our clothing, like members of the same bobsled team.

The attitude of intimacy and friendship persisted all morning. Whenever I chanced to cross paths with Tim or Wallace, there were joking remarks or side-of-the-mouth comments on the idiocies of some of the hired hands, who were rejoicing in the sudden day off by clowning in the barroom or shoving each other into the snow. Paul Cyr had come out two or three times to survey the snow as if he still might breast it in his ten-inch rubbers to get that final cord or two. Then he had settled to playing cards in the barroom, where Harold had joined him.

The busiest man in camp was Gus Lorry, who was making much work of the need to get ready for hauling, even though the teamsters were yet to be hired and there were sleds that lacked steel and whiffletrees on which the rings had not yet been sweated. Still Gus bustled about the hovel, hanging harness on pegs, finding straps that needed mending, and bemoaning the lack of rope. From time to time he would appear at the door of the hovel, where his head just cleared the drifted snow, and curse the fate that had half-imprisoned him. Actually what troubled him most, I am sure, was that the storm had made it impossible for him to trot panting up to Tim's office with word of new small disasters.

To me this surprise snowfall, which, by stealthily lapping all the visible world, had changed our shabby board-and-tar-paper settlement into a few infinitesimal lumps on a map, came as a blessing. Whether Eben Kendall had really concocted some plan to do me in or had simply been trying to learn what small secret I might have plucked out of the snow near Dunham's store, or if indeed that had really been Eben who had followed me down to the lakeshore — all this seemed of no more consequence now than if I had been wafted a continent away. And it was with a light heart that I completed some minor chores about the dooryard, helping Andy locate and retrieve the bolts of popple that had not yet been split, using Harold's borrowed snowshoes to tread down the path we wore between the office camp and the cookshack, freeing the outhouse door — the only one that swung outward — and shoveling the office steps completely bare of snow so ice would not form.

I was all the more dismayed, therefore, when, just before the dinner call, I spotted two black figures far out on the lake moving unmistakably closer. One figure seemed to ease along with a steady lifting and falling of both arms, as if he were equipped with wings; the other came ploddingly behind, obviously borne on snowshoes. The first figure had come close enough to grow a white face before I realized he was traveling on skis — implements I had seen worn only once before in this country. The dooryard by this time had been trodden into several beaten-down areas and pathways where men could move back and forth to the outhouse or between the office camp and the cookhouse, sometimes slumping unexpectedly right to the crotch, and getting snow up the sleeves and inside the mittens as they fought to climb to solid footing again. Then after two steps, they might sink again the length of a leg. But it was possible now to stand on the office steps or in the trodden-down bay outside the door of the barroom and watch the men come close. Before the two were near enough to recognize, half a dozen of the Dutchmen and the two French boys had come out to watch, like home folk gathered to greet an incoming ship. To me there was something frightening about these two shapes, materializing so mystically out of the vacant world and seeming to bear down directly upon me. I kept my eyes on them every second as they inched our way. Well, at least, I assured myself, neither one was Eben, for in his winter togs Eben looked like a grown bear; both these men were obviously slender and swift.

"Jim Kidder," said Harold. I agreed. Then, almost immediately, we realized it was not Jim Kidder at all. The man on snowshoes was more angular than Jim, and not so big, although he was lean as Jim was and just as long in the leg. It seemed that we must recognize one of the men before they came much closer, for they had obviously come down from the storehouse, where the denizens were all familiar to us as passengers on the same ship. But the nearer these men approached — one gliding well ahead, one chugging along behind at better than ordinary snowshoe speed, as if trying eternally to catch up — the more outlandish each one looked. "Son of a bitch," said Harold. "Who the hell *are* these guys? A couple of fucking prohibition agents?"

Prohibition agents? By God, the man on snowshoes was H. O. Stapleton! But whether he had come to smell out high wine in the hovel, to unmask some skulking Frenchman — or to cross-examine me, I dared not speculate.

"That's the sheriff," I told Harold.

"What's he after?"

"Search me," I murmured, already hushing my voice lest he overhear us from a quarter mile away.

"The other guy could be immigration," said Harold. "Only he'd have a uniform."

The man on skis turned out to be no immigration man at all, but just a boy, not so old as Harold and certainly not much older than myself. We watched his face take shape before us, a dark face with a long thin nose preceding it. Eyes so dark beneath his cap visor, they seemed black. He wore a red plaid "hunting suit" of knee britches and mackinaw, so bright and unsullied it seemed to have just been lifted from a box. He wore shiny black boots that reached partway up his calf, and red stockings that had been lightly sugared with snow. By the time he and Stapleton came close enough to hail us, the group of choppers gathered outside the barroom door had grown to a dozen, all staring gloomily at the approaching figures as if they were coming to take their jobs away.

To negotiate the high drifts near the shoreline, the boy on skis had to turn sideways and hitch himself up a step at a time like a very small child on a staircase. A few of the choppers dared to snicker at this. By the time the boy stood high on his pedestal of snow, Stapleton was pressing close behind him. The boy, apparently selecting Harold and me as members of the executive branch, directed his first greeting to us, a solemn and slightly breathless "Hi!" His face was red with cold but moisture shone on his temples. Everything about him proclaimed "rich boy" to me — his condescendingly "rough" clothing, his exotic skis, his dark-bearded face, so tenderly shorn that not a stray whisker had escaped alive. He could *only* have been the son of the boss. Yet neither Harold nor I realized this until he slid down off his snow pile and reached a hand out to us over the skis.

"I'm Gordon Curtis," he declared. Awkwardly we accepted his strong handshake, delivered in the collegiate manner of the day, with the elbow held high. We, each of us, mumbled our own names in turn and young Gordon made hardly a pretense of noting them. He turned and rendered a mechanical smile to the clustered lumberjacks, only a few of whom did more than briefly meet his eye and then look away. Stapleton mounted the snowdrift then, red-faced, with his nose leaking moisture. He sniffed deeply and told everyone howdy in an offhand manner, then slid down the snow pile to stand near Gordon Curtis. He grinned at me and offered me a special greeting.

"Howdy, young feller! They send you all the snow you need?" This indication that he might have come particularly to question me set my insides to squirming. I contrived nevertheless a sort of smile and returned his howdy. Harold meanwhile was paying obeisance to Gordon Curtis, smiling energetically and assuring Gordon that he had seen him once or twice at the company office in Berlin, New Hampshire. Gordon immediately pretended to remember Harold clearly, although he obviously could not have told us apart. It was against my own nature to cater to a child of my own age as if he were a grown man, so I stood mute, preparing only a shabby phrase or two to contribute if called on. The choppers, having stared their full at the boy whose easy life they were all financing, filed by ones and twos back into the barroom, and promptly broke into random laughter, muttering God knows what.

"Mr. Stapleton wanted me to bring him over here," said Gordon, using an artificially deep voice that was, I suppose, meant to indicate the importance of his mission. Stapleton, his weather-darkened face practically expressionless, winked at me.

"Didn't want to take no chance on getting lost," he said. Harold and I laughed but Gordon merely nodded as if it were perfectly natural that he, being a scion of the family that owned all the stumpage in sight, should have it given to him to find a way through wastes and wilderness that would leave lesser folk bewildered.

"Then Dad figured while I was here," Gordon went on. "Well, he thought I could sort of check a few things . . ."

"Picked a great time," said Stapleton pleasantly.

Gordon laughed. "We didn't exactly count on the snowstorm," he agreed. "It kind of caught us all with our pants down."

By this time, Tim had discovered the visitors and had come out on the steps of the office camp to size them up. While he would not have known Gordon Curtis even close by, he could not have failed to appraise the skis and the flagrant clothing that marked Gordon as a creature from a better world. Without taking the time to go back for his coat, in his greasy wool shirt and wide suspenders, Tim started down over the partly trodden path toward where we stood. He had taken no more than three steps when one stumpy leg plunged nearly to the knee and Tim, spread out like an ecstatic frog, dropped face first into the snow. It took him fifteen seconds or more to regain his balance. He dashed the oozing clots of snow from his face, wiped his burning hands on his shirtfront, set his face in a mirthless smile of welcome, and urged himself on. Again and again he sent one leg sounding the depth of the innocent snow and yet plunged on regardless, like a man charging desperately ashore. Harold and I took care not to laugh at the spectacle. Gordon Curtis smiled encouragement. But Stapleton lifted his small chin and laughed aloud.

"I God, Tim!" he shouted. "You're going to have a sled road all broke out there!"

When Tim reached us the effort had winded him so utterly that the words of welcome he had formed merely rasped in his throat. He kept bending forward to gulp fresh breath, even while he took Gordon's hand in welcome. Tim's face had turned a terrifying shade of purple. He held one hand to his side. Gordon Curtis, taking Tim's wet hand, seized his elbow too and held tight as if he feared the man might fall. But Tim pulled himself free, coughed wretchedly two or three times, and found himself able to protest that he was all right, indeed fine. He still did not know exactly who Gordon Curtis was and Harold had to tell him. Tim thereupon forced his painful smile wider still and squandered in even heartier greeting all the precious breath he had just recovered.

Tim then shook Stapleton's hand with appropriate vigor

but his grin shrank into a grimace as he looked Stapleton in the eye. He clung to Stapleton's hand an extra moment to ask him the question we all wanted the answer to:

"What brings you here in this weather?"

Stapleton, his pale blue eyes focused on some spot five miles across the snow, shook his head slightly and smiled.

"Oh, just nosing around," he said. "Just nosing around." He could not have selected any words better calculated to ignite panic in the hearts of most men in the camp. Even Harold, who surely had no sin on his soul more scarlet than a little unchurched fornication and commonplace traffic in illicit booze, exchanged with me a quick glance of fear. I think we both must have been moved at the same instant by an impulse we dared not yield to — the urge to duck into the barroom and warn Paul Cyr that if he *had* any booze within reach he'd better bury it deep. At the same time I was convinced in my own shrunken heart that Stapleton had some questions to ask me about my purpose in making that side trip to Dunham's store some ten days earlier. But Stapleton's first question was addressed to Tim:

"Who you got here for Frenchmen now?"

This was the question Stape had asked me at our last encounter and the answer Tim gave him was no different from mine — just big Paul Cyr, Robichaud, and a couple of kids. Stape nodded as if he had known all along what the answer would be — as indeed he must have. He made a faint sound in his closed mouth but said nothing.

"Well, Jesus Christmas," said Tim with sudden heartiness. "Let's not stand and freeze our balls off. Come in and have a cup of tea!"

"Go good," Stapleton agreed.

"Is there any coffee?" said Gordon Curtis.

"Best in the world!" said Tim. "Come in! Come in and set!"

Stapleton unbuckled his snowshoes with a small struggle, for his fingers had stiffened. Gordon Curtis got his skis unhitched in half the time and looked for a place to lean them. Harold hastened to receive them from Gordon's hand and to lay them slantwise on the snow-cushioned eaves of the barroom. Then we all waited for Stapleton, who carefully stuck his worn showshoes tailfirst in the drift. After some brief maneuvering and waving of hands, Gordon was persuaded to proceed us all through the barroom door. Three lamps were burning in the barroom, making the place as dingy and foreboding as an opium den. A few of the men lay stretched on their bunks, while the rest had all gathered within the embracing breath of the box stove. Paul Cyr, towering above all the squat Dutchmen, eyed our procession solemnly. With his deep eyes and fierce whiskers, and his wool cap plopped rakishly far back on his head so that a black forelock fell over his brows, he might have been the hunted leader of a den of assassins. But when Tim and the visitors had passed, Paul released his nearly toothless grin at Harold and me. He nodded his head back and forth to the rhythm of the clumping footsteps of the small parade. "Hup-hup! Hup-hup!" he sang, half under his breath. "You in da army now!" I stopped for less than a second to lay my hand on Paul's arm. "Stapleton, the sheriff," I whispered. All mirth fled instantly from Paul's face. "JEEZ-a Chri'," he breathed, glancing instinctively behind him to see if his escape were cut off. But Harold and I pushed on to stay up with the others and we dared not glance back. The communicating door between barroom and cookhouse banged shut after us and from the barroom side there came a scuffling about, as if everyone in the barroom had been suddenly pressed to make top speed for the outhouse.

We all five — Gordon Curtis, Stapleton, Tim, Harold,

and I — sat at one end of the table farthest from the barroom door, where cups and spoons had already been set out along with sugar and a pitcher of evaporated milk. Andy, the cookee, bore a bucket-sized teapot over from the stove and set it in front of Tim, while the cook stood near the stove, holding a long metal spoon and watching us all uneasily. It was less than an hour to dinnertime.

"You got any of your good coffee ready, cook?" Tim called out. "This is the boss's son. We got to treat him right."

"All ready," said the cook, in a feeble voice. He directed a nervous smile to Gordon, while Andy, before the cook could lay a hand on it, lifted the coffeepot from the stove and awarded it to us.

Tim had to stand up for the job but he managed to transfer coffee from the elephantine pot into Gordon Curtis's cup without spilling more than a small puddle. While Gordon and Stapleton noisily ingested tiny sips from their steaming cups, Tim, his eyes crinkled half shut from the effort, set out to charm Gordon by asking him if he brought this weather with him. Harold and I were so intent on trying to divine what might be taking place behind the door to the barroom that we hardly heard what else was offered on this subject, or any other. When the commotion inside the barroom reached a crescendo, we both flinched and exchanged the flick of a glance. But Stapleton, devotedly cuddling his cup of hot tea, seemed not to have heard. And even if he had he could no more have unriddled the sense of the stumbling and banging about than we could. Did they *all* have booze to hide?

When the visitors had taken their fill of hot coffee and tea, we all, at Tim's behest, paraded one after one to the office camp, along the now squeakily packed and soiled footpath that was sunk more than eighteen inches into the snow. In the office the little iron box stove had been sturdily chugging out waves of warmth, unattended; now the flimsy camp was snug as a nursery. Gordon Curtis shed his jacket and cap here, revealing a surprisingly frail physique, shoulders that protruded in small knobs above his chest, arms that seemed about the same thickness from wrists to shoulders. His hands were lean and wiry and his handshake had been as vigorous as any man's. But denuded of his shell this way, with only a wool shirt and a thin tight sweater to disguise him, he looked like a child. A very rich child. Gordon seemed to sense my suddenly deepened interest in his person, for he looked directly at me in an almost plaintive way. In the segment of a moment before he could compose his face into a pleasant smile, it wore the wretched look of a little boy who has lost track of all his playmates.

"You a college man?" he asked me. Not having yet got used to acknowledging I was a man, I found myself blushing and trying to wet my throat with saliva.

"Not exactly," I replied. But before I could undertake to report the semester I had put in at Brown, the door opened noisily and Wallace brought in a slice of winter. Squeezing one more man into the tiny room made us all move an inch closer and look about for spare corners. Harold hitched himself back onto his bunk, stretching out with his head resting on his clasped hands. But Wallace did not want a seat. He pulled his mittens off and ritually rubbed his hands over the stove, his mild blue eyes expectant as he made note of the visitors.

"Oh," said Tim, half standing, "Wallace, this is Gordon Curtis. *The* Gordon Curtis." He made a joke out of the last remark by laughing briefly, nodding all the while and grinning his stiff grin at Gordon. "My brother, Wallace," he told him. Gordon stood up to shake hands, again in his high-elbow style.

118

"You're the junior partner then," said Gordon.

"Well," said Wallace softly, "I owe half the bills." Whereupon Gordon and all of us laughed more heartily than we needed to. Tim gestured at Stapleton.

"Stape you know," he said. Stape held his place and nodded five or six times, smiling up at Wallace.

"Oh, I know Stape," said Wallace. "Don't know anything *bad* about him." Again everyone laughed dutifully. Stapleton made a mock-skeptical face. "Don't imagine you know a hell of a lot that's *good* either," he said.

"Well, you manage to keep your tracks pretty well covered." This brought another, and louder, outburst of laughter from everyone. In this wash of goodwill it was almost possible for me to believe that Stapleton had walked in three miles over the drifts merely to make his manners, being in the neighborhood. But my skin still prickled faintly at his presence.

"You two staying to dinner?" Wallace inquired. And before anyone could respond, he added flatly, "Course you are!"

Stapleton and Gordon merely looked at each other as if seeking permission.

"Well," said Gordon, "whatever Mr. Stapleton . . ."

"Never passed up a free feed yet," said Stape.

"Well, now," said Wallace, "why don't I take the young fellow over with me? I know Ma would want to meet him. And I think she's got a lamb stew on the stove. Something smelled awful good."

Gordon looked rather woefully about at all of us, who had obviously not been included in the invitation.

"Better go along," said Stapleton. " I don't know what Elizabeth has on the stove, but whatever it is I bet it's a damn sight better'n any woods cook ever set out."

"Oh, we don't do so *bad*, Stape," Tim protested. But I

dared not look at Harold, lest the whole room read the glance he would give me.

"Well, come along then . . . ah . . . mind if I call you Gordon?" Wallace said. "We don't use mister a hell of a lot around here. Kind of get out of the habit."

"Oh, hell," said the boy, "everybody calls me that. I *like* it." He worked himself back into his mackinaw and fitted his cap square on his head, so his ears were comforted. He looked once more about the room with an expression somewhat short of joyous. "Then I'll see you all later," he said. We each offered a polite murmur and he followed Wallace out into the snow, with Wallace explaining that he would go first to show Gordon where it was safe to set his feet.

There was a deep silence when the door closed. All of us seemed to be waiting for Stapleton to announce his purpose in coming, but I dared not look directly at him, nor could I fish up any phrase to start the talk off in a tame direction. Tim finally slapped both hands on his knees and said, "Well . . ." It was as if he had given the signal for dinner: The cookee's call came at once — a halfhearted cry, for it was seldom indeed, on a weekday, that the crew took their noon meal in camp. Stapleton stood up and poked about our little wooden sink for a sliver of soap.

"Better wash my forward feet," said he. He poured a dollop of water from the steaming kettle into the basin and domesticated it with half a dipper of cold water from the pail. He sloshed his hands briefly in the water and dried them on the soiled towel that hung on a nail, then studiously emptied the basin into the sink and watched the suds flee down the dark little drain hole into the freezing snow-walled cavern under the camp. I watched his every move as if he were preparing for a hanging. No one spoke. None of the rest of us ever washed up before dinner, unless we had been doing work that soiled our bare hands. Harold got halfway out of his bunk and waited, as I did, for Tim to lead the way. But Tim still sat on his chair, watching Stapleton's effort to find the nail from which the basin had been hanging. And he was still sitting there when we heard the hurried crunching of feet on the snow outside. Andy burst the door open, his naturally gray face pink from cold or excitement.

"Hey, Tim!" he gasped. "You'd better get down here!" Tim did not move from his chair. He pursed his lips slightly and opened his eyes wide, in an expression of utter doubt.

"What's the difficulty?" he inquired.

"Jesus!" said Andy. "I don't know what's going on! The crew won't come to dinner! Seems like they want to fight the cook! Or something."

"Fight the cook, for Christ's sake?" Tim got up from his chair.

"Well, I mean, they're calling him names from behind the door. I think they're all hotter'n hell!"

Stapleton reached over and hit me on the sleeve. "Come on, young feller," he said. "Let's see what this is all about."

I had no urge whatever to discover, or to help Stapleton discover, what had been going on in the barroom. But I jumped to my feet and followed along all the same, it still being my nature to obey my betters. Tim followed more slowly, taking the time to attire himself in coat and hat. Andy waited to go down with Tim. Stapleton was first in the door of the cookshack. We had heard stamping and yells and laughter from the barroom as soon as we started across the dooryard but they hardly seemed wild enough to start men running for help. Even in the cookshack, the noise from behind the wall of the barroom offered no menace. But there were only four choppers at the tables — two Dutchmen and the two French boys — all of whom looked up at us as we entered, with almost identical expressions of foreboding. They all dropped their glances quickly and concentrated on their plates. The food had already been set out on the table in pannikins and platters, where it steamed, unspoken for. The cook looked up, white-faced, from moving pots about on the stove and came a few steps toward us.

"You got revolution on your hands?" said Stapleton.

"Oh, they're like a bunch of goddamn kids!" the cook muttered. "They get a whiff of high wine and there's no holding them. I'll chuck the goddamn dinner in the stove in five more minutes!"

Tim came in now and his face instantly began to redden when he saw the empty tables. He strode halfway toward the barroom door and let out a throaty yell:

"Come and get it or you don't eat!"

The door was pulled open about half the length of an axe handle but no face appeared. Out of the shadows behind it one deep voice bellowed: "Stick it in your ass already!" This was followed by whoops and squeals of laughter, some nearly high-pitched enough to be girlish. The door slammed tight. Tim took another step toward the door, as if he were about to push it open. But he stopped, turned to glare at the four innocent choppers who were hunched close over their plates, then turned back toward us. His face had grown nearly purple again and bubbles of saliva appeared at the corners of his mouth.

"That's the last fuckin meal any of *those* bastards eat in *this* camp!" Tim snarled. "And they'll pay for it, too!" He laid hold of the platter nearest to hand, where it sat forlorn on the marred oilcloth, and gave it half a shove, as if he meant to sling it and its burden of still faintly smoking gray hash into oblivion. But it moved just a few inches, abandoning two or three damp crumbs along the way. Then Tim started toward us, his face still gorged with blood.

"Well, let's *us* eat dinner!" he growled, as if by devouring a healthy share of the grub we might commit an act of revenge on the crew. But Stapleton, with half a smile on his face, did not follow Tim to a seat. He took me by the arm and drew me along toward the barroom door.

"Come on, son," said he. "Let's see what's going on in here." A tiny arrow pierced my stomach, for I assumed at once that Stapleton suspected I at least had guilty knowledge of what was afoot. Else why drag me to the scene of the crime? But I went along without protest, indeed without even a sound, except the suddenly accelerated beating of my heart, which was imparted to my ear alone. We slipped into the barroom without opening the door more than enough to allow our slim bodies to pass, Stapleton first, with me at his heels. The lamplight here hardly diminished the gloom, and the windows were darkened by the drifted snow. The moment we entered, two men lifted cups to their lips and gulped the contents, tipping their heads back to capture the final trickle. Most of the faces were veiled in semidarkness. Only Paul Cyr's face stood out clearly, aflame in the light of the nearest lamp, his mouth opened in silent laughter or surprise. The whoops and the shouting stopped the instant Stapleton and I were well inside the room. Then the far door opened, and in the silent explosion of white light, two Dutchmen ducked outside and flung the door shut behind them. Other men retreated into the caverns of the double-decker bunks. There was some muttering and snickering and one young man still sat at the foot of his bunk with his head in his hands. Otherwise, complete sobriety seemed to have laid hold of the entire company, as if Stapleton had cast a spell. He was a sinister-looking figure in the dim light, long-legged and leathery-skinned, with a way of lifting his eyes so he seemed to be focusing at some spot just beyond the immediate scene, like a hunter watching for movement far ahead. He was a skinny man but his hands were large, and knotted from labor. His small intent face, sharp as a mink's, gave an impression of great ferocity. One or two of the Dutchmen were just drunk enough to meet his glance directly for a time, in a sulkily defiant manner. But they looked away quickly and stumbled out of his path. Halfway through the room, Stapleton stopped for a moment and looked deep into the corners, just as if he could see what might have been tucked away in total darkness. He did not speak to anyone until he reached the outer door, then he turned to me and barely murmured a few words that sounded like "Let's go." Go where? I wondered. But I stayed at his heels as we went out the door and followed the soiled pathway to the outhouse.

The Dutchmen who had escaped had gone only a few strides up this path and at the sight of us they looked rather frantically about, as if contemplating plunging off into the drifts. Instead, they both stepped into knee-deep snow just off the pathway and faced us glumly, seeming to expect at the least to be laid hold of and dragged back into the camp. When it became clear that Stapleton was going right on past them, they began to giggle, fixing their gaze on me. "Too late already," one of them snorted in what was meant for a whisper. Then they both laughed out loud, slapping each other on the chest and shoulders.

Stapleton chugged ahead, in his bent-kneed stride, without offering the two men so much as a nod. The wind had lifted suddenly, dodging through the dooryard in sudden eddies, now from the west, now from the north, and flicking dry snow in our faces like sand. Stapleton made straight for the outhouse, where a few torn papers fluttered free on the snow, although most of the mess had been gentled and laid deep to rest by the foot or more of snow that had sifted in under the shelter. Stapleton gave one brief glance into the pit, where late deposits showed up black at the bottom of miniature wells in the snowdrift. He looked all about in the nearby thicket, among the bowed blackberry branches, each one adorned with a two-inch ridge of snow that clung through some magic to even the daintiest twigs. What the hell he was seeking I could not guess, for no footprints led off into the brush, nor had any of the berry bushes been shaken bare by someone's plunging through. But Stapleton lifted his chin suddenly, then reached up a mittened hand, shook snow from the branch of a small fir, and broke it off at the trunk. With this he reached out under the bushes, about twice the reach of his arm, and hooked a metal handle that was hardly visible above the snow. He drew the prize back and lifted up an oblong metal tin, like a maple syrup tin, with a screw-cap opening at one upper corner. The cap, however, was missing.

"Here we be!" Stapleton exclaimed quietly. He held the can up before me, then sniffed at the opening. "High wine surer'n hell," he murmured. He turned the can up and shook it. A few flecks of snow fell free but not a drop of liquid drained from the opening. Stapleton laughed almost happily and looked at me with his eyes bright. "They done away with the evidence in a hell of a hurry!" he said. "No laying hands on it now!"

"I guess not," I said. There was a sudden letting go in my throat and I laughed somewhat foolishly. Stapleton gave the can a fling and watched it perch atop an arch of snow-laden brush. Then he turned back to me and looked me straight in the eye, with a change of expression that startled me.

"What give you the idea Eben done that shooting?" he demanded.

I could feel the blood surge into my neck and face, then

drain quickly back. He could not have surprised me more utterly if he had struck me with his fist. My chin trembled when I tried to speak.

"I — I — n-never . . ." I began. But my God, I asked myself. *Could* I have told Jenny's suspicion to anyone? I was certain I had not. "I didn't say that," I managed at last to reply. Stapleton continued to look me right in the eye. His gaze was so sharp and steady, his light blue eyes so unforgiving that they seemed to chill my soul.

"Never had no suspicion?" he insisted.

"Well, somebody said . . ." I gasped, and instantly realized that I had said too much.

"Who said?"

"Well, nobody." I could hardly breathe from embarrassment and had to pull in air through my mouth.

"Jenny White," said Stapleton, smiling suddenly. He had one gold tooth that barely showed at the corner of his smile.

"It wasn't Jenny!" I blurted.

"It was nobody," said Stapleton mildly. He shook his head at me and struck me on the arm. "You're a good boy. Let's see if we can make out some kind of lunch down here."

"Well, it wasn't Jenny!" I insisted, dismayed to think I had betrayed her.

"I ain't going to *say*, even if it *was*," said Stapleton. "Rest easy."

He set off back to the barroom and I tagged behind. Most of the choppers had filed into the cookshack now and were making a pretense of eating. One plate had fallen to the floor and distributed its burden of hash and bread crust at the unheeding feet of two of the choppers. Nearly every face turned toward us as we entered but only one man dared call out:

"Find anything good, clerk?" Then everyone laughed

raucously except the sober few who dared not even smile. I shook my head.

"Not a drop," I said. At this, one young Dutchman close by slapped me merrily on the arm. Far down the table, Fred Aulenbach, his long blond hair matted as if someone had emptied a cup over him, suddenly grew solemn, stared hard at his plate, his mouth protruding, then struggled up off the bench, put one hand to his mouth, and plunged for the door. Stapleton had to stop to let the young man stumble past. Stape shook his head.

"Well," he said, in a quiet tone. "It *will* make you sick." At the "office" end of the far table, only Harold sat, nibbling at a molasses cookie, Tim's spot being already vacant.

"Tim lost his appetite," said Harold, as we climbed in beside him on the bench.

"Don't wonder," said Stapleton. He reached out for the platter of dingy hash and pushed about a third of it onto his plate. I poured myself a cup of tea.

Stapleton required no more than ten minutes to take his fill of hash and bread, upon which he had poured molasses. He drank about half a dipper of black tea, then got up to carry his soiled wares to the sink.

"I'll be on my way," he said. "You tell the young feller good-bye. Believe I can make it without him." He winked at me as he said this. I finished my tea and followed him out the door. Harold joined us out in the snow, to watch Stapleton buckle his snowshoes on. His only good-bye to us then was a series of nods. We stood shivering, coatless, in the sharpening wind to see him set off across the lake. A cluster of choppers appeared briefly in the door to make sure Stapleton had truly taken his leave, then they banged the door shut and shared their rejoicing. The wind, sweeping dry snow off the lake surface, to pile it in deepening drifts along the drifted shore, quickly wrapped Stapleton's plodding figure in a swirling veil. He moved along straight for the distant landing, as if someone had gone before him and drawn a line for him to follow. As he moved farther into the distance, his figure would sometimes disappear completely, to appear again as a sort of half-realized ghost, just a head and a body, without legs, or just half a body. Then he was gone altogether and Harold and I, squeezing ourselves to keep the final ounces of heat in our ribs, scuttled back to the office camp.

Tim lay on his bunk, pale-faced and frowning. He barely lifted his head to learn who we were, then fell back and whispered a few curses.

"You sick?" said Harold. Tim rolled his head a little.

"Started to have kind of a falling-out spell," he muttered. "Fuckin coffee."

"Stuff'll kill you," said Harold. "Quicker'n booze." Tim, peering from under swollen lids, got his eye on me.

"What the hell did old single-tit want?"

"Stapleton?" I hesitated for hardly a second. "Oh, he was looking for booze."

"Find any?"

"Just an empty high wine can, up by the outhouse. What they did was drink it all up just to get rid of it. Not even a drop in the can."

Tim muttered a few curses and then sat partway up.

"What a fuckin comedy! Three miles in *this* weather just to sniff out a little high wine!"

"That's what they pay him for," said Harold.

"Where the hell is he?" Tim said throatily. "Gone back?"

"He's halfway to Oquossoc," I said. Tim sank back on the bed and put one hand over his eyes.

"He take the kid with him?"

"No. He never even told him good-bye."

Tim rolled his head back and forth.

"And what the hell the kid's snooping around for, I don't know. Just because we can make a dollar or two where the other contractors *starve*."

"Oh, I don't think he knows anything about the business," said Harold. "They just send him up here to get him out in the fresh air. He ain't been too well."

Tim's voice had diminished almost to a whisper. But he spoke through his nose, feebly imitating Gordon Curtis.

"His *Dad* wanted him to *check* on something."

"Oh, he just said that," Harold protested. Tim grunted faintly. Harold, who was standing near the small window, ducked down to see if Gordon might be coming back.

"Well, Jesus to Jesus!" he exclaimed. "Will you look who's coming now! The worse the weather gets the more goddamn traffic we see!" He flung the door open, admitting a hatful of dry snow that whisked in and lightly sprinkled the floor. "Well, you picked a fine goddamn day to come calling!" he shouted into the outdoors. I moved over to the door and saw Jim Kidder plodding toward us on snowshoes.

"Figured she was going to blow," he yelled at us, "so I come to find my wandering boy. Wouldn't want him to get lost."

"Will you close the fuckin door!" Tim croaked behind us.

"Come in! Come in!" Harold cried. "He'll be back here any minute. Come in and warm your ass!" He pushed the door most of the way closed, then opened it again when Jim had shucked his snowshoes.

Jim's wool hat carried a thick frosting of snow and there was snow in his eyebrows. He stamped snow off his boots and trousers, and took his hat off to slap it against his leg.

"Merry Christmas!" he shouted, as if he were still across

the yard from us. He grimaced when he saw Tim stretched out.

"The old hoss down again?" he cried. "What they been giving you to drink, Tim?" Tim sat up and tried to put on his company smile, but merely managed to twist his mouth slightly.

"Just off my feed a little," he said in a scratchy voice.

"All that rich food!" said Jim, and shook the cabin with his laugh. "Them meatballs and that red flannel hash! Right, Bob?"

"Coffee," said Tim.

"I tasted that coffee you fellers put out," Jim chuckled. "Wonder you can still draw a breath. What you done with my boy? You feed him some of that brew?"

"He'll be along," said Tim. "Just finished his dinner." Tim lay back again and once more set his hand over his eyes.

"Hey, you're really sick," said Kidder, in a suddenly subdued voice. "You got something to take?"

"All kinds of stuff," Tim murmured. "I'll be all right, after I get a little shut-eye."

"Well, Jesus," said Kidder, "let's leave the man in peace. I want to take a look in the hovel anyway to see what your livestock looks like. Guess they're all fit to start hauling."

"They will be," Tim whispered, without opening his eyes.

"You go ahead," said Harold, who had got his boots unlaced. "I'm going to stretch the frame out for a few seconds myself."

I decked myself in warm coat and wool hat and followed Jim out to the hovel. Following the deep tracks where Tim had wallowed along an hour earlier, we slumped several times in drifts above our knees. The hovel was half hidden by a drift that had been sculptured almost to a knife edge by the wind, which still stole handsful of snow from the top ridge and flung it in swirling patterns toward the roof. I had no stomach for being questioned further about Stapleton's errand; I was half distracted still from fretting over what might now befall Jenny White. But goddamn it, I told myself, I never did tell him it was Jenny.

Exactly as I feared, Jim Kidder, once we had labored up over the drift and found shelter in the hovel, turned to face me.

"What the hell did old Stape want with you?" he demanded.

"With me?" I shook my head. "He was looking for booze."

"Oh, shit a goddamn," said Jim. "He never come three miles in this kind of going to find six dollars worth of high wine. He was asking all about you before he started out."

I felt my ready blush creeping up to my ears again.

"What'd he want to know about me?"

"Oh, like where you come from, what you did nights, you screw many women, where you bought your liquor."

"Balls," said I, for Jim was grinning the whole width of his face.

"No," said Jim. "He just asked how long you been in these parts, if I knew what kind of gink you were. I told him I just knew you from seeing you here."

"He just went looking for high wine here," I said. "The crew was all half sloughed at dinnertime."

Before Jim could question me further, Gus Lorry, swathed in a whole trunkful of misfit clothing, so he might have been a man, a ghost, or even a gorilla, slid down from atop the drift and knocked two pounds of snow into the hovel. He was panting, as usual, and his eyes seemed to start from his head.

"What's up?" he gasped. "Anything wrong?"

Jim laughed. "We just been sizing up some of your stock for the glue factory," he said. "You going to *haul* with these critters?" Jim looked about the dismal reaches of the hovel and waved one hand at the serried rumps.

"Well, sure!" said Gus. "Don't have no choice!"

"You're going to have to grain the hell out of them first," said Jim. "Old Wallace'll scream when he sees what it's going to take to build any steam in this lot." Jim wandered down between the rows of stalls. One or two of the beasts shifted and nickered when he came close, or turned as far as their halters would allow, to roll a white eye at the stranger. I came a pace or two behind, but there was nothing new for me to see in the gaunt array of bony hindquarters. The fragrance of the droppings stung eyes and nostrils.

Jim shrugged after reviewing the stalls. He opened the grain bin and made note of what was stored there.

"Don't forget to get them bridle-chain hitches on the sleds," he said to Gus. "You got what metal you need?"

Gus flung his two hands, enshrouded in tattered mittens, high above his head, and waved them as if he had been attacked by hornets.

"Don't talk to me about bridle chains!" he yelled. "That ain't no part of my job! I don't know the first goddamn thing about what's going on with the blacksmithing! And I don't want to know! I got my hands full right here with this barnful of critters to feed!"

Jim laughed indulgently. "Well, you'll lose a horse or two on that mountain if them bridle chains ain't on the sleds on the downgrade. Right, Bob?"

"Don't ask me," I said.

"Well, you sluice one of them sleds with six cords of

pulpwood in the rack and she'll pick up your fucking team and make dog meat of them. I've seen some quite messes in this country, back along!"

"That ain't none of my worry!" Gus proclaimed. "That's for the teamsters to look out for!"

Jim winked at me.

"You going to teaming, Bob? Or what the hell they going to set you to?"

"I don't know," said I. Nor did I. I was no teamster, I was sure of that.

"Maybe they'll make you a road monkey," said Jim. "How are you at spreading hay?"

"Oh, I can do that if I have to."

"Less'n you get tired of Tim's horseshit."

"I'm tired of that already."

"How'd you like to go to cutting ice? I got a little contract over here, I could work you in."

"Anything to get out of here," I said.

Gus, his tattered mittens dangling scarecrow style, his devastated wool hat drooping almost over his eyes, stood with his mouth open.

"Don't let Wallace get wind of you quitting," he said in a sort of stage whisper. "Without you got some paper says what your pay is, by Jesus, he'll give you short scale. I seen him do it a hundred times."

"One way to beat that," said Jim. "Ask him for a month's pay to send home for Christmas, like the Dutchmen do. That way, you'll get the rate down in black and white at least. Right, Gus?"

"Right!" said Gus, solemn as a preacher. "That's what I'd do, by Jesus. You get it written down and they can't beat you. Not that they won't try any way they can. I know that pair from way back!"

We were still standing in the hovel when the hail came from the office camp: "KID—DAHR!"

"My wandering boy is back," said Jim, and we filed out of the drift to flounder back to camp.

"Crying out loud!" Gordon Curtis exclaimed, when we entered the cabin. "You didn't have to come *get* me! I know my way back."

Jim nodded. "You may know it, but I ain't so goddamn sure you could find it, the way it's blowing now. You might wind up in the brush. There's goddamn little left of the track you and Stapleton made. Times out there, you can't see three feet ahead of you."

"Well, if it got too bad, I could stay right here," said Gordon.

This remark brought a look of dismay to Tim's face; he and Wallace exchanged one quick glance, then Wallace hastened to caulk up the sudden crack in the conversation by pretending to have to clear his throat. He spat very carefully into the sink, expelling a gob of saliva as daintily as a girl.

"Glad to have you," he said. "Plenty of room!" There was indeed an empty bunk above Harold, where the cook had temporarily bedded down. Tim rubbed his hand over his forehead and eyes, reclaiming his sick spell. Then he managed to murmur: "Fix you right up here if you need to."

"Christ Jesus, no!" said Kidder. "It's fixing to come down hard again and you sons of bitches may be holed up here until March! I got to get this boy home safe. If I leave him here, he may take to using bad language."

"Oh, nuts," Gordon laughed. "They won't teach me anything I don't know."

We all contributed some laughter, even Tim, who felt able to say, in a reasonably solicitous tone: "Might be better to get going at that. There *could* be another storm making up, the way the wind has shifted."

134

"Not too goddamn much daylight left either," said Jim. "We better get them Christly barrel staves on your feet and get you moving."

Gordon, who was obviously no more eager to spend the night with us than Tim and Wallace were to have him, agreed quickly and began to button himself into his clothes. He offered elaborate good-byes to Tim and Wallace and charged Wallace to bear his *million* thanks once more to Mrs. McCormick for that really *great* lamb stew. Harold and I walked with Gordon and Kidder down to the drifted lakeshore and stood beating our mittened hands together while Gordon fastened his skis.

"I still can't figure what Stapleton wanted with you," said Kidder, squinting happily at me. "He didn't trek over here tell you hello."

"Maybe he wants to make Bob a deputy," said Harold.

"More likely he wants to line him up for that skinny daughter of his," Harold laughed. "You ever see his daughter?"

"I didn't even know he had one," said I.

"Nice enough girl, I guess," said Harold. "But Jesus, she's got no more tits than a pickerel! May be five foot eleven, and don't much weigh more'n a handful of horseshit. I don't know where the hell a man would find room for his cock. Be like driving a wedge in edgeways of a shingle!" He let go his wild laugh, which the wind seized and sent echoing about the whole clearing.

Gordon, like the rest of us, joined in, shaking his head and jabbing one hand toward Jim.

"Talk about bad language!" he said. "What the hell is he teaching me?"

The door to the barroom opened and Paul Cyr stepped partway out, grinning, ready to join in the joke.

"Here's the scoundrel," said Jim, who was tightening his

snowshoe clips. "Just missed getting hauled down to Farmington, hey, Paul? Old Stape is hot on your trail."

"Not me," said Paul. "I don' do nothin'."

"Hell you don't," said Jim. But Gordon had his skis all fastened now and turned his bright face to us to perform his farewells. He gave my hand, which I had hastily bared to match his, a special shake. "I hope I get to see you again, when I'm up. We've got to talk about college."

I was startled by the earnestness with which his clear brown eyes looked deep into mine, the way a lover might look, and by the note of entreaty in his voice. I was not used to being courted by the rich. "Oh, sure," I mumbled, "sure," looking away as I spoke. Jim had already started off toward the dimly visible mountain to the north and Gordon, nodding once more to all of us, turned and slid away on Jim's trail. We watched the lifting snow enwrap them as they moved off, one behind the other.

"So they gone," said Paul.

Harold turned and eyed Paul intently. "You got rid of that evidence in one hell of a hurry."

"Sure," said Paul. "I sell it."

"You *sold* it?"

Paul's whiskers parted to show the deep red cavern of his mouth. He laughed happily at us. "Righ! Righ!" He shouted. "I sell the sunnamabeetch! Every fuckin Dutchman in camp, she's owe me a dollar!"

136

❧ 5 ❧

HAULING began when the snow was so deep that it would hide a horse right to his ears. The teamster, drawing two trailing loads of bundled pulp sticks to break out a road, would ride on the second bundle, far enough behind so as not to slide up on the horse's rump if the beast should balk at the steepness of a sudden ram down. Once a good road had been broken on the downgrade, the loaded racks would be drawn down more directly to the shore, to be emptied out on the frozen lake.

Men would stand then along the road, which would gradually turn slick and shiny, and would fork hay out on the sled track, to help keep the loads from "sluicing" — running away, that is, impelled by their own weight, and carrying the team with them. Spreading the hay was an endless job, for each sled would gather up the hay into windrows as it slid over it, then it would all have to be spread out over the track more or less evenly to help brake the team that followed. On the steepest grade the pace of the sled had to be slowed by means of a length of bridle chain hooked around underneath the sled iron in what was known as a bear-trap hitch, which could be knocked loose with one blow of an axe, so the sled could run freely once it reached level ground. On many jobs where the grade was dangerously steep, snub lines were used to hold back on the rack full of pulp as it coasted down behind the team. These were long ropes, hitched to the back sled and wound around a tree trunk. A wooden brake would be driven into the trunk just above the coil of rope, so that it could be pushed down against the rope to keep it from unwinding too rapidly. The man who stood at the end of the rope and worked the brake, or fed the snubbed rope to the tree trunk by hand, had to stay constantly alert lest the rope escape and unreel around the trunk in a smoking fury, allowing a rack of pulp to smash into the hapless team and send it to destruction.

The McCormicks would never have trusted me with any such job and I wanted none of it anyway, for I would have lived in terror with that burden on my conscience. Instead, I was set to loading pulp at the top of the grade, yanking the sticks off a pile with a needle-sharp pulp hook, and tossing them up into the rack to be piled snugly together. This was a frantic enough task for me, who had never wielded a pulp hook more than casually, and who was given to stumbling and sliding about in my desperate efforts to keep pace with the teamsters. Actually the job of yanking the half-frozen sticks off the pile and heaving them to the bed of the rack was twice the job of piling them and most of the teamsters acknowledged this. But in every such crew there was always at least one hand ready to seize any chance to flaunt a spoonful of authority. One skinny, shock-haired young man always drove up toward me wearing a smirk, as if he were savoring in advance his delight at putting me down. He always pretended to bear my slowness and awkwardness with extreme patience — patience that was being tried to its outer limits. Most people in the woods converse in loud shouts anyway but this lad, whose name was Jack Scott, invariably bellowed out his comments to me so that everyone within a quarter mile could hear them. Jack was no favorite of anyone, for he was given to outbragging every man in camp. There was no hunting ground, no matter how remote, nor any pond or stream, in whatever hidden corner of nowhere it might lie, that he had not fished in or trod upon, no game he had not bagged, no willing lady he had not himself enjoyed, nor any boss he had not taken the measure of. Whatever strange accident might have befallen any of his associates, the very like, if not something far stranger, had happened to Jack. And all the marvels of nature that other men had set eyes upon, Jack had seen them all before they did, or soon after.

So Jack, despite his unvoiced appeal to all the other teamsters to come share his distress at my failings, never won any sympathy from his mates. Indeed, one of the Rangeley teamsters, Walter Hamm, invited Jack one time to try swapping jobs with me and Jack allowed solemnly that By God no man was going to load *his* team but himself, being as *he* was the man had to *answer* for it if they was anything wrong. At this Walter chuckled darkly; I was comforted thereby and even dared look Jack in the eye and second the invitation.

Poor Jack, as it happened, had something wrong to answer for before the second week of hauling was out. It was a bright cold morning, and the road was all ice as he brought his smoking team up to gather his first load. I watched the nodding horses pull the empty rack up abreast of the pulp pile, and I took care to avoid Jack's eye, for I knew he would be offering me his look of weary resignation before he even had to deal with me. My glance fell instead on the icy chains that trailed alongside the forward sleds and I waited to see Jack hop out and hitch them in place. Instead, Jack merely clambered into the bed of the rack and let go a yell:

"Well, Jesus Christ, clerk! Ain't you done dreaming? You'd oughta get your sleeping done on your *own* time!"

Because I had heard this identical jibe from him more times than I wanted to count, it hardly raised a welt on my vanity.

"You going to hitch the bridle chains first?" I asked him.

Jack did look mildly abashed, for the better part of a second, but he recovered himself at once.

"Don't you fret yourself about my bridle chains, young feller!" he shouted. "You handle your end of the contract and I'll be able to handle mine a damn sight better!" I had already dug the point of my pulp hook into the nearest stick and I began to swing the rough bolts of wood one at a time up onto the rack, where Jack, still protesting to the wide woodland that I would do my own job better if I would be pleased to butt out of his, grabbed them up and arranged them in rows.

My part of the job always left me drawing in deep painful drafts of zero air, convinced that I could not pick up one more stick, nor even hold the pulp hook another two seconds at the end of my aching arm. Unable, or at least unwilling in Jack's presence, to pause long enough to comfort my numb face, I worked with tears running unchecked clear down over my chin and with my nose building a puddle on my upper lip. So when I saw the load start off at last, while Jack flailed his whip and shouted, "Git up in that goddamn collar!" I saw, and yet I did not see, that the bridle chains still trailed unhitched alongside the sled. That is, I observed that they had not been hitched but I was too done in to tell myself about it.

It was not until the sled had mounted the slight rise and started on the steep downgrade that the meaning of the loose chains registered.

"Hey!" I hollered. "Hey! The chains!"

But by this time Jack too had begun to yell.

"Whoa! Whoa!" he screamed, as he felt the overloaded sleds pick up speed beneath him. I began to run after the team, as if I might catch hold of the rack and by main strength keep it from sluicing. But it was already moving faster than I could and Jack's wild yells had become wordless sounds of terror. Yet he clung to the suddenly slackened reins and waved his useless whip even as the loaded rack, now completely out of control, drove into the rumps of the terrified team and swept them on.

Harry Ernst, who had entered the main road just two or three rods ahead with his own team, turned back to see the source of the racket. He made one vain and frantic effort to whip up his horses to clear the way, then leapt out into the snow, where he sank to his waist. His team plunged partway into the snowbank and stuck there. Jack, too, then leapt for safety, still hanging tight to his whip. The thundering load of pulp drove the two horses straight into the rack ahead and seemed to crush them into the jumbled wood. A shaft splintered, with a crack like a felled tree. Pulp was spilled out of the rack and hid itself in the snow. The horses screamed like human creatures in awful pain and struggled wildly to get free. The near horse, as he tossed his head, flung a stream of blood into the snow, where it scattered in bright red gobbets, like spilled marbles.

Two dozen men gathered from everywhere, some floundering through the drifts, some scrambling along the icy way. Tim McCormick, his face suddenly gray, pounded up the slope, mouth wide-open and eyes aflame. He joined the others in the desperate effort to cut the trapped horses loose. The near horse, free of his harness and pawing wildly at the dumped sticks of pulp, stumbled at last into the snow, where he sank to his belly and stood there mournfully wheezing, his muzzle an oozing red mass, like a blood-soaked sponge. The other horse seemed to be struggling vainly to free himself from a trap. Even when the tugs and belly strap had been unbuckled, he kept pawing with one hoof at the pulp sticks beneath him, while one front leg was held fast. Harry Ernst found a way through the scattered pulpwood and splintered poles to get at the anchored hoof and he let out a yell of dismay.

"Jesus Holy Christ!" he hollered. "See what this poor fuckin critter done to hisself! He's driven that shaft right through his fuckin hoof!" Harry clambered out to find an axe and set out to cut the horse free. When the horse was led out at last, it wore a long splinter of wood, three inches thick, stuck right through the cleft in its front hoof and protruding six inches in front. "Never touched no flesh!" Harry exclaimed breathlessly. "Didn't cut him a bit!"

Foolishly, with his open mouth pouring white smoke, Tim McCormick seized hold of the splinter and tried to yank it by main strength free of the horse's hoof. But it stuck as if it had grown there.

"You ain't going to get that out that way," Harry told him. "Only way is to cut into the hoof. You may have to pare it down some."

But Tim kept trying, gasping out curses in a sort of fiendish whisper, for he was nearly spent from running up the slope. After one final yank, which almost cost him his mittens, Tim sat right down on the ice and rocked himself back and forth, eyes shut, like a man knocked down in a fight. Breath scraped in and out of his throat but he had none to spare for talking. Harry stood over him.

"You feel all right?" he said. Tim waved one hand weakly, then shook his head and spat.

"Out of breath," he gasped.

The whole crew by this time had gathered about the scene, to help move the wreckage off the road, and to rescue the buried pulpwood from the drifts. Jack Scott, wearing clots of snow all down his front, from his hat to his trousers, and still carrying his useless whip, stumbled down toward where Tim was sitting. Jack was pounding the air with his whip and shaking his head to the same rhythm.

"I swear to *God!*" he was crying to any ear that would attend him. "I swear to *God*, I had them bridle chains *hitched!* Must be someone *knocked* them. Or . . ." He slid a glance toward me, as he appraised his chances of shifting the blame to my back. "I swear to *God*, they was *hitched* when I was *loading!*" He seemed very near to weeping. His mouth kept twisting as if he were in pain.

Harry Ernst stood up from where he had been bending over Tim and turned to face Jack Scott.

"You're a fuckin liar already," said Harry, as calmly as if he were reporting on the weather. "Clerk was yelling to you about the chains when you come over the rise."

Jack's expression of studied bewilderment dissolved as if he had been struck in the face. His chin went limp and his mouth hung open. Only a faint sound came forth as his breath tried to catch its balance: "I . . . I . . ."

Tim had got himself to his feet at last. He wiped spittle from his lips and made a gesture with one mittened hand as if he were tapping Jack, who was yards away, in the chest.

"Never mind that," Tim gasped. "You . . . you just . . . get the hell down to camp and let the clerk figure your time."

Jack had his mouth back in working order now. His face went very white. "Jesus Christ Almighty, Tim!" he wailed. "You got to give me a chance!"

But Tim had used all the breath he could spare. He clung to a corner of the devastated rack and kept his eyes closed.

He did not seem to hear Jack at all and eventually Jack started back to camp, with me a few strides behind, and Jack explaining to me and to the empty woods and to himself how unfairly he had been dealt with. In the office camp, while I fished up the book to figure his time, Jack kept his tear-filled eyes fixed on my face and talked as if he and I were both privy to the fact that it had been he — Jack Scott — and none other, who had warned of just such a disaster as this.

"I don't know how many times I told Tim . . . Hell, you must have heard me say it Christ knows how many times . . . I told him to get rid of that Christly bear-trap hitch! Christ! I don't *like* that goddamn hitch. You seen what can happen, coming over that rise when I knocked into some fuckin stump or something and that hitch just come undone! You seen that yourself! You know Jack Scott don't load up without he's got them bridle chains hitched! You ever know me to load up without I had them bridle chains hitched?"

I dared not add to Jack's misery by answering him truthfully, so I pretended to find some scribble I had to squint at to decipher. But Jack hardly left room for an answer. "I like that old-*time* hitch," he declared breathlessly, "where the teamster hung right on to the bridle chain. Then if anything went wrong you knowed it right off! These goddamn bear-trap hitches, they've lost more hosses! I've seen it! I've seen more good hosses lost from using them fuckin bear-trap hitches. Not only in this country but over in Magalloway, back along. Save *time!* Shit! You ever try to knock a bear-trap hitch loose when she's all iced up under the sled? Why I've spent as much as twenty minutes trying to knock one of them bastards loose with an axe. Had to crawl right under the sled to get at her! I told them what could happen. But oh, no! They know better. Some of these big brains down to Berlin. They ever try driving a two-sled rig in the woods? Like hell! But you can't tell *them!* Well, now you seen what happens!"

Having satisfied himself that his reputation had been repaired, Jack had become nearly his old self again, no longer bent half across the desk to beseech my sympathy, but sprawled in his chair and waving one red hand shoulder-high to decorate his speech. I had made out the company order in his name and I pushed over the slip on which I had done my figuring, pointing out the deductions for tobacco and other trifles. Jack, half smiling, with an almost scornful expression, hardly took time to read them but picked up his order and folded it small to fit it into his purse.

"Well, now they seen what can happen," he said, with his voice back in his chest once more. "Maybe next time they'll *listen* when somebody that knows his business tries to tell them something. They won't take the word of a teamster! All right, now they lost a hoss!"

Jack would not wait until the men had returned to camp and someone might have given him the ride back to Oquossoc to which he was entitled. He obviously had no stomach for touting to any other teamster the virtues of the old-time hitch. He'd *walk* out, by Christ!

"Why, shit," he told me. "I've *swum* acrost bigger puddles than this."

And off he went, floundering through thigh-deep drifts to reach the roadway the tote team from the storehouse had broken out to travel down the lake to the village. The cook and the cookee both came out of the cookshack to look after Jack, unbelieving.

"Hey!" the cook cried. "Whyn't you wait till one of the other fellers gets back to give you a lift?" But the cook was not used to shouting — indeed he talked hardly at all — and his thin voice seemed to fade into a whisper before it reached halfway out to Jack. So we watched him until at last he found the road, shook snow off his trousers, then set off southward, with his straw suitcase tied to his shoulders like a backpack, a shabby and forlorn little figure on the wide white lake. I could not even rejoice at his going.

To get the brutal splinter from the wagon shaft out of the horse's hoof, they finally had to cut the whole hoof away, paring it down to a nubbin, so the poor creature hobbled on a short leg. Gus Lorry led the horse down the road and across the yard to the hovel, with Tim walking behind.

"You'd best get him off in the woods there," Tim called to Gus, "away from the hovel as far as you can go and shoot him out there. Get him where he won't stink up the yard."

Gus came to a halt and the poor horse stood on three legs right at Gus's shoulder, his ears alert, his big head turned back to eye Tim, almost as if he were about to enter a plea for his life.

"I'd just as soon keep him in the hovel," said Gus. "That hoof'll grow. I could tend him."

"A hell of a lot good that would do us," Tim replied wearily. "He'll be eating his goddamn head off all winter, taking up space, and not pulling a fuckin stick."

"I'd tend him," said Gus. "I won't give him no grain. What's the difference, if you're going to shoot him? Why not give him to me?"

"Who the hell'll pay for his keep?" Tim snarled. He shook his head to indicate he had borne all the idiocy he could bear. "*You* going to pay board for the son of a bitch, for Christ's sake?"

"I *will!*" said Gus, moving a step toward him. "I'll pay for what feed he uses."

Tim studied Gus for a few seconds in silence.

"Cost you half a dollar a day," he said.

Gus did not speak at once. I stood in the half-open door

146

of the office, looking from one man to the other. Half a dollar was a hell of a price for what little hay an idle horse might need. It was half of what Gus was earning. I could not believe he would divide his wage with a horse. But Gus allowed that the price suited him fine and he made off to the hovel, leading the horse, as jaunty as could be. Tim looked after him for a moment, then looked up at me, almost guiltily, as if I had caught him in some misdeed. He set his face in a glum expression and made his way slowly up to the camp.

"You can't expect a man to board a goddamn hoss for *nothing*," he declared. Then he stretched himself on his bunk, not even removing his hat, and gave out a sound that was both a sigh and a moan.

"What a fuckin day!" he muttered.

"Jack's gone," I told him. "I got his figures here if you want to see them." Tim made no sound at all for a second or two beyond the noise of his heavy breathing. Then he opened his eyes and blinked at me.

"He *walk* out, for Christ's sake?"

I nodded.

"Just as well," Tim whispered. "Save me from booting his ass."

I ducked out then and left him to his nap, while I visited the lame horse in the hovel. Gus greeted me by slapping me gleefully on the arm, his mouth wide open in silent laughter.

"I got me a *hoss!*" he told me in a sort of half whisper. "I got me a damn good *hoss!*" His delight at becoming the legitimate owner of a property so massive and so regal was almost more than he could contain. He grabbed my hand in his own grimy fist and shook it as if he had just come into an inheritance. It seemed almost certain that this poor crippled horse was the first treasure Gus had ever been named the owner of, beyond an extra pair of brad boots or a two-dollar axe. As I stood there and helped him admire the sweating animal, Gus kept grabbing my arm to give it one more shake.

Tim came down to supper that night still gray in the face, his mouth drawn down and his eyes weary. He ate hardly at all, said nothing, held his pannikin of tea in both hands, as if for comfort, and carried most of his supper away to dump it in the pail by the sink. I watched him all the way out the door, doubtful now if I wanted to stick to my resolve to ask tonight for a reckoning of my pay. There were but ten days left until I was due to join Jim Kidder on the ice-cutting job and I wanted some sort of interval to elapse between my asking for a payday and my suddenly taking my leave. Otherwise, as Gus had warned me, I might be awarded short scale. So it had to be today or tomorrow or the day after that, and Tim was likely to be just as far off his feed at any of those times as he was now. It would be Wallace I had to reckon with anyway, so I trudged up through the snow to the little office camp, feeling as uneasy as I had often felt when I had to make application for a job or appear before the school principal to make excuses for an absence.

But Tim lay unmoving on his bunk when I came in and did not even lift his head to make note of me. I looked into the stove to see if it needed wood, and then I sat in silence. Tim and I seemed to hear at the same instant the sibilant steady breath of Wallace's gas lantern as he moved up through the dark yard, for as I turned to face the door, Tim pulled himself erect and began to rub the sleep out of his face and eyes. Wallace, as always, opened the door gently, nodded his head as he looked at one of us and then the other, and offered no more greeting than his standard "Unh-hunh," uttered without opening his lips. He set his lantern down on the floor, took a seat on the bench by Tim, and warmed his fingers ceremoniously over the lantern. After a full minute, during which none of us spoke a word, Wallace shifted his feet and sighed.

"Well," he said, in a voice hardly louder than a whisper, "let's see what we've got in here."

He moved then into the small wangan room, lugging his lantern, and by the light began to check over the daybook.

"You got that Scott joker paid off," he said in a tone half questioning and half a declaration of fact.

"Paid him off and got him the hell out," said Tim throatily. "Walked out, pack and all."

"Walked out?" I heard Wallace move his chair back uneasily, ready to cope with trouble. "You *offered* him a ride." Again his remark was only partly a question.

"Never saw him after Bob gave him his time. I'd have offered him a kick in the balls."

"He wouldn't take a ride," I put in hastily. "Even the cook tried to get him to wait. He said the hell with it. He wouldn't listen to either of us." I was not sure if I had really offered him a ride or if I had simply questioned his haste, but it was no time for me to own myself delinquent in any respect.

"Well, you can't *force* a man to ride," said Wallace, largely to himself. "All you can do is offer." He flipped another page in the book. "Never even said good-bye," he murmured. "Now I call that pretty damn rude." I recognized this for a joke and laughed dutifully. It seemed an excellent time now for me to make my request. I idled casually to the office door and looked in on Wallace, who sat with his reading glasses well down on his nose.

"By the way, Wallace," said I, vainly trying to sound offhand, "I was wondering if you could let me have what I got coming to me, so I could send some money off home."

Wallace had set his lantern beside him on the little shelf that served as a desk. He had to squint across the light, lifting his head to see over his glasses. He sucked in his cheeks and seemed to fix his gaze on the bridge of my nose. After a time he put his hands down flat in the open book and stared at them for a minute or more, working his mouth almost as if he were tasting something.

"Supposing I was to let you have a check for fifty dollars," he said.

None of my rehearsings of this scene, in odd times through previous weeks, had included any such response from Wallace. I think I stood with my mouth open for a second or two before I could utter a word. It was exactly such an offer that had prompted young Ernst, weeks earlier, to declare that that was "not what I asked for when I come in here." And for a desperate few seconds those words were all I could find that seemed to fit. Wallace kept staring at me over the lamp.

"If that will fill the bill," he prompted me, in his mild little voice.

"Well," I replied, my voice on the very verge of squeaking, "I kind of wanted to get paid up. I mean, I needed more than that. My mother . . ." My mother had not really come into it at all. I realized I had begun to sound like a small boy begging for a handout. My face was burning red.

Wallace clapped the daybook closed and took off his glasses.

"Well, that kind of puts a different face on it," he said sharply. "You want to get paid up. You're heading out then."

"I wasn't heading out." My voice had become very weak indeed and I had to swallow to wet my throat. "Just wanted as much as I could get."

Wallace nodded and raised his voice to direct it at Tim.

"What do you think we ought to pay this man?" he called. "The kind of work he's done?"

Tim shuffled into the glow from the lantern, his red face set in its most unpleasant expression, a sort of prissy, contemplative look, as if he were sampling something not quite to his taste. He did not look at me at all.

"We can get service like *that* for a dollar a day." He spoke in a nasal tone that made his words doubly offensive.

"What the hell are you talking about?" I cried without thinking. I was alarmed to find that I was on the edge of tears. "I was making three dollars . . ." My voice broke then and I could not proceed.

"You quit that job to come here?"

"Well, no, but I never would have . . ."

"Well, give him an extra quarter a day if he's not satisfied," said Tim.

"If you're going to leave us right in the middle of things," said Wallace, paying Tim no heed, "walk right off when we're at a busy time, you can't expect top pay. Supposing we was to find there was some mistakes in your accounts here" (he tapped the book as he said this) "and we'd be out twenty or thirty dollars on some one of these jokers that's been all scaled up and paid off and gone home. And supposing you was off on a new job cutting ice somewhere . . ."

This remark chilled me, for of course he had put his stubby finger right on the truth I had been so stubbornly treasuring. But his next words exploded in my ear.

"Not that you'd be worth any more than a dollar a day to Jim Kidder on a job like that."

Embarrassment scalded me so now that the gas lantern seemed suddenly to have flared up into a flame that half blinded me and seared my cheeks. I just checked myself from saying: "Who the hell told you that?" for it would be doubly embarrassing to try to argue that it was not true. Instead I let my embarrassment boil into anger.

"I heard what a stingy pair of bastards you were," I declared, in a choked voice. "I never believed it until now." Of course I had believed it all along and I realized very quickly that my store of insults and angry phrases was scanty indeed. I had managed to sting them both all the same, and they hastened to accuse me of merely repeating accusations marketed by some of the "jealous skunks" who infested Oquossoc and had never owned a pot to piss in.

The only retort I could contrive then was to declare solemnly that my information had come from some "fine people" — a phrase that sounded fatuous enough even to my ears and that prompted Tim and Wallace to emit almost identical bursts of artificial laughter. The childishness of this exchange merely added to my misery. I had no real stomach for name-calling, so I sulkily urged them to make out my order, scanty as it might be, and let me get shed of this place in the early morning.

Wallace then reckoned my time in his painful scribble. With exaggerated politeness he laid the paper before me (it looked like some third-grader's homework) to point out how he had arrived at his final figure. My eyes were really too wet to make out the numbers he was indicating. I accepted the order silently and folded it into a small square so it would fit into the shabby wallet I carried in my hip pocket. I would cash it tomorrow at Mackenzie's in Oquossoc.

I had an impulse to plunge immediately out of the little cabin into the dark. But where would I go? The cook extended no welcome to late callers and I had no appetite for the barroom, where most of the men were still strangers to me. Sunk in a welter of self-pity, I relished no company but my own.

And now, I mourned to myself, there was no job to go to either. For how could I face Jim Kidder, after his having betrayed me to the McCormicks? On the matter of "friendship," at that age, I was solemn as a priest. I don't recall that my own readiness to set claims of friendship above my personal comfort and convenience, or business advancement, had ever been tried. But I had learned the tenets of that religion from dozens of stories I had wallowed in and speeches that had bated my breath, all glorifying loyalty to one's buddies as the most soldierly of virtues and the one signal emblem of manliness, to be prized above the square jaw and the stiff upper lip.

It is not that I for the smallest moment dared even privately suspect Jim Kidder of being short of manliness. It is simply that I knew, through this event, that I had been deluding myself with the notion that I was included among big Jim's true friends.

Wallace and Tim had lingered for a moment in the wangan room, whence their whispers did not reach me with any meaning left. For want of another retreat I stretched myself, boots and all, upon the bunk that had once been Harold's and I stared into the dark corner where the yellow lamplight was lost. Oh, Jesus, I told myself, what a hell of a fix I've got into now! I could feel the sobs trying to form themselves in my mouth. Being out of a job in that era, you must understand, was nearly the ultimate tragedy for any poor man who had never been able to lay up a dollar against hard times. When the paycheck was gone, there were no government funds to stave off beggary. And for a young man who had flaunted his decision to head into the wilds to find his fortune, there was the added shame of creeping back dead broke to face the neighbors he had forsworn.

I tortured myself by living again the scene at the hovel when Jim Kidder had first offered me the job. But the face

I saw first in the vision was that of Gus Lorry, wide-eyed, foolish, hair on end. And I heard his warning me, in that don't-tell-a-soul manner of his, that I had better take care that the McCormicks not give me short scale. But Jesus! The thought pierced my insides like a steel sliver. It was *Gus* who must have told the McCormicks what I planned! Gus! That miserable, hairy old son of a bitch! To think that *he* had run tattling to the McCormicks after all the secret gripings we had shared!

The realization that it had not been Kidder at all suffused me so with relief that I hardly had space for anger at Gus. It was the sort of contemptible thing I told myself that you had to expect from a worthless old prick like that. I would not waste breath ever talking again to the bastard, I decided, for I had not yet lived long enough to understand that if I cut off relations with everyone who repeated what I had told them in "confidence" I would live a lonely life indeed. Anyway, it was done, and perhaps just as well. For now I could openly take off for my new job with no pretense of leaving on some sudden impulse. Except, I realized immediately, that my new job was not ready for me. Well, then, I would simply stay in Oquossoc somewhere — anywhere but the Dead Rat — until Kidder needed me, which would be only ten days from now. Perhaps Jenny, I comforted myself vaguely, would rent me a room. . . .

Wallace by that time was ready to return across the brook. He grunted his standard farewell and carried his hissing light through the cabin, lifting me out of my reverie with the stark glare of the lantern, then leaving the cabin in sudden twilight, as if a curtain had been drawn. I sat up and began to rid myself of my boots and trousers. Tim puttered about the little table, where he had laid his magazine, then he too made ready for bed, his breath coming in quiet gasps, as if he had just climbed a set of stairs. I did not undress as extensively as I usually did; I left my socks and shirt on, woodsman-style, and slid beneath the greasy covers, with my back to Tim and the light. Tim was ready in a very few seconds.

"You through with the light?" he inquired in a scratchy tone.

"Uh-huh," I grunted and he blew the flame out. He seemed to sit very still for a while in the dark before he took to bed. I held my breath for fear he might have something more to say. But he was silent. The boards in his bed creaked and he groaned faintly as he settled himself. I tried to fix my mind on what I might do tomorrow to get myself to Oquossoc, to find a bed, to find a meal. . . . But I was asleep before I could sort out a single vision.

The "turnout" call of Jim Sullivan, the new cookee, awoke me, as Andy's always had, coming faint and faraway from the barroom. When he called again, from outside the cookhouse door, I was wide awake and sitting up.

There was usually a fifteen-minute lapse between his first call and the call for breakfast, time to snuggle for a last few seconds before braving the cold floor. But this morning I wanted to get clear of the cabin, if I could, before Tim was stirring. He had always been the last one out of bed but sometimes sat coughing, clearing his throat and trying to make conversation while Harold and I started the fire and got ourselves washed and dressed.

I scratched a match on the stove top to light the lamp and in the flare of the match I saw a faint glitter from Tim's partly opened eyes. The old bastard is wide awake, and watching me, I noted. I avoided looking at him then but concentrated on getting the kindling lit and standing close to the stove to still my shivering. I had put in cedar enough to start the fire crackling within seconds. I set the washbasin on the stove and broke the thin ice in the water bucket to dip out just enough water to wash with. But I had no patience to wait until the water was warmed. I doused my fingers in the water, rubbed them on the soap, and made a few ritualistic passes at my face and hands before wiping off the chilly water with the gray towel.

There was a flow of heat now from the stove, enough to gratify my chilled flesh as I stood beside it. I fitted a stick of silver birch into the fire, set the damper, and heard the cry for breakfast.

I looked over then to see if Tim was starting. He had not moved at all. His mouth stood partly open, the way old men's mouths do when they are helplessly asleep. His eyes were dark slits, with the moist pupils barely showing. Then a sudden realization came to me. Except for the heat of the fire, there was not a sound in the cabin. Of course I had been distantly aware of that all along, but now I made conscious note of the fact that I could not hear Tim's breathing.

Fear seemed to lay an icy finger across my stomach, sending a tingle all through me. I moved a half step closer, not daring really to bend over Tim, lest he sit up suddenly and snarl contempt into my face. But he was *not* breathing. There was no movement of his belly or chest, not the faintest whisper from mouth or nose. More than that, there was a dirty gray color to the skin of his face, as if the last trickle of blood had left it. Hardly daring to breathe myself now, I reached out and took hold of Tim's naked wrist with finger and thumb. It was cold as oilcloth. I had to swallow hard to get voice enough to speak.

"Tim!" I said sharply. His eyes did not even flicker. They seemed each to hold, between the half-shut lids, a smear of something shiny as glue.

"Holy Jesus Christ!" I whispered. For nearly a minute, I was unable to make myself finish dressing. Savoring the animal fear that constricted my heart, I stood transfixed and watched the dead man — the very first I had ever seen. But I knew of course that I would have to go at once to summon Wallace, or anyone else who could come confirm that Tim was truly dead. Excitement began to enrich my blood now, for I knew that I alone would be bringing tidings of Tim's death to all the camp, striking everyone silent. As I shoved my arms into the sleeves of my coat, I was ashamed to find myself relishing this role and I made a serious effort to compose my features into an expression of grief.

It was black dark outside, for the sky was clouded over. We kept a lantern eternally lighted in the wangan as protection against freezing the apples; I hastened back to borrow this and turned the wick up as I eased myself down the icy steps of the cabin. A dim halo of yellow light attended me across the tumbled dooryard. The pathway Wallace had worn in his journeys back and forth across the brook to his own cabin was firm enough to hold me without snowshoes.

There was light in the window of Wallace's cabin and the smell of fresh wood smoke drifted over the brook. But when I rapped on the cabin door, there was deep silence for several seconds. Then Wallace pulled the door back and faced me. His eyes widened at the sight of me and he took a quick step away, apparently fearing I had come seeking some wild satisfaction for the deal I had been awarded. His tousled hair and flushed face made him seem years younger.

"It's Tim!" I told him. "You'd better come over!"

Wallace's mouth fell open and he seemed not to have understood what I said. His wife, wrapped in a man's bathrobe that folded double over her bosom, stepped into the light. Without her glasses and with her scanty hair astraggle, she looked like a stranger.

"One of his spells?" she asked me hoarsely.

"I don't think he's even breathing."

The fear and woe that contorted Wallace's face then really startled me.

"Son of a bitch to Holy Jesus!" The words burst out from a throat that seemed constricted in anguish. Wallace grabbed up his coat. Then, realizing suddenly that he had no boots on, he dropped the coat and began a search for his rubbers. He pulled them on hastily, tied the laces around his ankles without lacing them up, and plunged for the door. I had intended to lead him back with the lantern but he shoved past me and scuttled along the path without missing a step. He reached the office camp ten or twelve strides ahead of me and when I came in, he was holding the lamp up to view Tim's body. There was no missing, in the bright light, the pallor of the frozen face. Wallace set the lamp back on the table and turned toward me, glanced at me once, then turned back to the bed. Profanity poured from his mouth like a stream of tears.

"Of all the fucked-up rotten deals! This is what the poor son of a bitch . . . this is what he gets! Instead of taking some of these cocksuckers around here! . . . The fuckin lousy pricks that run this fuckin world . . . A man who broke his ass all his life . . . Oh, my stinking rotten Jesus!" Wallace was actually weeping as the curses fell from his tongue. Then of a sudden he was still, as if someone had grabbed him and stopped his mouth. He stood looking down on his dead brother, his elder self, without moving. He stepped toward the bunk after a moment and reached out one hand, only partially unfolded, to set the tips of his hard fingers for just the briefest touch against the dead cheek, in a gesture almost unbelievably tender. Wallace sat down then, in the nearby chair, and looked down at nothing.

"I could stick around awhile," I whispered. "In case you

need some help to . . . ." I didn't know what he might need help with nor what I dared try to provide.

Wallace shook his head without looking up. He took in a deep breath and seemed to shake himself free from his grief. He looked me straight in the eye.

"Well, if you could spare another day. We'd make it up to you. They's a few things we could use a hand on." He was whispering too now, suddenly respectful in the presence of the dead.

"That's all right," said I. "I don't have anyplace special to go. I could go get the doctor. . . ."

Wallace could not restrain himself from twisting his mouth in the manner he always used to indicate his impatience with stupidity.

"Doctor can't do much for him now! But if you *could* get over to the storehouse and get word to Kendall . . ."

Kendall! Well, of course, he would have to come here, being the only undertaker within fifty miles. But Jesus! To have him right here in camp, where I was friendless and the same as alone. I stood with my mouth open, unable to devise an escape.

"You wouldn't need to *walk* it," Wallace whispered. "You could have Gus hitch up a sled."

"No, no," I said. "It wasn't that." But I did not dare try to explain what it was. "It was just . . . well, I thought you might need a hand here first." How could I say out loud that I was afraid the undertaker might try to shoot me dead?

Wallace stood up and looked again at Tim.

"No," he said, no longer whispering but talking in the sort of tone you might use in church, "I'm going to leave the poor fellow just as he is. . . ." He stopped and tightened his lips, obviously repressing a sob. He had to swallow before he could speak again. "You'd help us an awful lot if you

159

could get word to Kendall. And they might have a rig at the storehouse we could use to . . ." Again he stopped and drew in a deep breath. "To . . . take him out where . . ." He was unable to say another word and had to turn away from me altogether.

"I'll start up as soon as it gets light," I assured him. But it had turned light already and the lamp glowed feebly against the gray window. I left Wallace in the cabin and went down to bring the news to the cookshack. The first man in my path, when I entered the cookshack, was the cook himself. I set one hand on his arm and told him in a low voice that Tim was dead. The cook stood frozen, with the smoking pot lid in his hand.

"Dead?" he gasped.

"Who's dead?" said Jim Sullivan, who had come up toward us grinning, carrying the fat teapot, ready for a joke. When I told him Tim had died, he too stood motionless and open-mouthed, gaping into my face. He put the teapot gently back on the stove and carried the word then to the men at the tables, all of whom, one by one, gave over the sedulous scraping of their plates and stared up at me in wonder. In the whole crew there was not a single man who bore the least love for Tim McCormick, and there were surely a dozen who had at some time openly wished him dead. Now every face was solemn and a few were aghast. Why just last night he had sat there . . . just yesterday he had stood beside them in the snow . . . just hours ago they heard him calling crankily across the yard. . . . How *could* he be dead then? Was he really *dead?*

I suppose most of them, like myself, must have harbored some guilt for the hatred they had borne old Tim and the villainous fate they had often, in everybody's hearing, consigned him to. And like myself everyone in the shack must have shivered a little to feel these icy feathers brush so close.

It took me nearly an hour to snowshoe out to the tote road and make my way, snowshoes in hand, up to the storehouse, a bleak red building set squarely at the northernmost reach of the lake. I found Jim Kidder in the cramped office in the building next to the storehouse, where a telephone with a hand crank on it was affixed to the wall. Jim greeted me with glad surprise, and did not notice my own reluctance to respond to his laughter with my own. Ignoring his suggestion that I must have made my escape in the dark, I told him flatly that Tim was dead. The swiftness with which this sobered big Jim brought me a sudden guilty satisfaction, as if I had been momentarily promoted to a rank almost equal to Jim's.

"His heart?" Jim asked me in a hushed tone.

"I guess so," said I. "He died in his sleep."

"He never looked good," said Jim. "Lately he's been looking worse. Always pale around the gills. Jesus Christ Almighty, though. I never thought they'd *kill* that old bird."

"I'm supposed to call Kendall," I said. "But I just as soon you do it, if you don't mind. I don't hit it off with that guy."

"Kendall? He got his eye on you? Hell, you'll put him in *his* grave."

"It's just that we don't hit it off. I just as soon not have any dealing with him."

Jim regarded me steadily.

"Kendall ain't going to touch the body without some doctor gives the okay. He's got to have a paper guarantees the guy is dead."

"He's dead all right."

"Kendall's not going to take your word, nor Wallace's either. Don't know as I blame him."

Kidder was grinning again and, feeling no real sorrow now, nor any further urge to maintain a doleful air, I laughed too.

"The old boy was really broken up, though," I said. "He really cried. Honest to God cried."

"Well, sure. They were awful close. More than most brothers. Awful close." Jim contributed a few solemn shakes of his head. "Well, I'll call old Doc Morton. As long as he knows for *sure* the guy is dead, he'll come hopping to get that five-dollar bill. Even so, it's going to be the middle of the forenoon before he gets here. We'll have to send the team down to fetch him from the landing. Might as well pack old Kendall in the same time. You want to ride down?"

"Christ, no. I better get back and give Wallace a hand. I said I'd stick around. I got done last night but I told him I'd wait over a day if he wanted."

Jim frowned at me.

"You got done? You ain't forgot about that ice job?"

"No. But I got thrown out quicker than I expected, that's all. I asked for my pay and they paid me off."

Jim laughed gently.

"Oh, you can't get ahead of those birds. Smarter than shithouse rats, those two." He made a wry face. "Only one of them now. Don't imagine he'll take another woods contract, with Tim gone. But what the hell are *you* going to do meanwhile?"

"Oh, I'll make out. Maybe I can rent a room from Jenny White."

This time Jim Kidder really let out his full roaring laugh.

"You watch out for old Jenny!" he yelled. "She'll put you through your paces! She ain't got but a double bed in that

162

shack of hers. You may have to bed down in the hovel to protect your virtue!"

"Jenny doesn't scare me," I told him. But truthfully I was mildly repelled at the thought of sharing a bed with a woman that old.

"Maybe I'll have to come chaperone you," said Jim. "Else old Stapleton'll come down and take you for living in sin. You and ol' Jen. You'll need someone to spell you after a day or two anyhow!"

"No chance," said I.

Jim took to the phone then and after some shouting managed to raise Dr. Morton in Rangeley. Reaching Kendall was easier, for he answered the first ring and accepted the call without surprise. I suspected he had made a rough estimate of just when almost any neighbor past fifty, far or near, might need his services. Hearing the man's rumbling voice on the telephone — for it carried half across the room — set me to fretting again about discovering ways to stay clear of him. If there was some way I could hide up here, without shame . . . But I had made my promise to Wallace. And there was no way I could ask shelter from Kidder without inviting his ridicule. I could perhaps just keep out of sight until they had started the body on its journey.

It was past eight o'clock by the time they had the rig hitched up and started on its two-hour journey to the landing. Jim had chosen a long pung, with removable seats, one that had not been used for several seasons and that left rust streaks on the snow when it was drawn out from the cluttered shed. It had been a passenger vehicle mostly, wore faded gilt decorations along its side, and so seemed more fitting for this errand than the battered and barren wagon body that was in almost daily use to bring freight up from the village.

I rode down to a point opposite the bleak camp, where I

climbed on the wagon track and slid my boots into the snowshoe clips. With only a short way to go, I did not buckle the straps and consequently had to shuffle along in my own tracks as if I were wearing bedroom slippers.

No one moved in the camp yard. Smoke came from all the roof jacks and there was the faint rumble of moving feet in the barroom. Obviously the teamsters had taken a holiday in honor of Tim's death. The horses, set wondering by this extra Sunday, were stamping and nickering in the hovel. As I made my way to the office camp, I realized it was not a place where I really wanted to be, sharing the warmth of the small stove with a corpse. But I could not turn away now without owning myself afraid, and so I stuck my snowshoes tailfirst in the drift beside the stairs and plodded purposefully up to the door. I stopped an extra moment or two to knock snow from my boots and this brought Wallace to open the door for me.

"Mmmmm," said he, stepping back at once to make way. I pulled off my hat the moment I came in and noticed at once that Tim lay covered now with a spotless white sheet that disguised the shape of his body.

"Kidder's sent for the doctor," I told Wallace in a low voice, as if there were a sleeper in the room. Then I quickly responded to the irritation showing in Wallace's face: "Kendall won't . . . *do* anything unless he gets a say-so from the doctor. He says." I added that last so there would be no argument over whether Kidder knew his ass from his elbow on such matters. Wallace's frown only deepened.

"Everybody has to get his," he murmured. "Cocksuckers. They work together." He sat down then in the chair nearest Tim's bed and looked solemnly at the floor, as he had probably been doing for an hour or more. Tim's worn boots stood side by side halfway under the bench that ran alongside the bunk. They still faithfully repeated the shape of Tim's ankles and stubby feet, except that the toes turned up slightly from holding no flesh to keep them flat. There was something so ineffably forlorn about these beloved boots that had comforted Tim all fall and winter, that I felt my throat grow tight. Tim's Sunday suit, undoubtedly rescued from some safe closet in Wallace's cabin, and stiffly draped on a wooden hanger, hung now from a nail on the far wall, ready for the undertaker to carry off and fit for the final time to Tim's arms and legs. Dwelling on small sad facts like these I lay down on my bunk and stared mournfully at the unpainted boards of the roof, where small drafts stirred the strings of cobweb that still clung there. The wind picked up from time to time and whistled faintly at the corners of the camp. Snow sifted feebly against the windows. Wallace sat without a sound, except an occasional drawing in of a deep breath and the complaining of the chair when he moved his weight to a new position.

Mrs. McCormick awakened me when she opened the camp door and let the bright light in. I sat up instantly, seized by a fear that Kendall had walked in and taken me unawares.

"They've come," said Mrs. McCormick, in the same hushed tone we had all used inside the camp. "They," to me, meant Kendall and his crew. I scrambled at once to get into my outdoor clothes and find some escape. But Wallace had stood up and he and his wife waited together in the open doorway, so I would have had to shove them aside to get away. I had hardly had time to think this thought when the heavy feet sounded on the steps, and someone knocked heavy boots on the planking to rid them of snow. Kendall seemed to fill the whole doorway, blocking the light so he could hardly see inside. He got his wool stocking cap off with one grab of his left hand, tucked it in his mackinaw pocket, and took Mrs. McCormick's right hand in both of his.

"I was awful sorry to hear about this," he growled.

Mrs. McCormick whispered some acknowledgment that no one could hear and Kendall immediately fixed his solemn and practiced frown on Wallace and got hold of Wallace's hand. He had not removed his mittens.

"I was *awful* sorry," he repeated, in a voice much louder than we had dared use. I could not stop myself then from imagining how Tim might have mimicked Kendall's unctuous tone and how he and Wallace would have snorted together to recall it. By this time Kendall's eyes had accommodated to the dim light and he started toward me, to assure me too that he was awful, awful sorry. But he recognized me almost at once and his face darkened in a scowl so ferocious that I could only back away. He turned away too, chewing a corner of his lip, and suddenly so disconcerted that he completely lost track of what his next phrase should be.

Dr. Morton, a plump man whose round face was so swathed in wool that he looked like an ad for a toothache remedy, had pushed himself forward then to reach the corpse first. His cheeks shone like ripe apples. He looked owlishly from one of us to another.

"Now, if the family will *excuse* me," he said.

This meant, of course, that we would all have to leave Tim to the ultimate mercies of doctor and undertaker. All the same, Wallace moved over to the bed and fumbled with the sheet long enough to find the lifeless hand that lay beneath it. He closed his fingers briefly on Tim's hand and then started out, bending his head and pretending he was having difficulty fitting his clumsy hat on. Before any of us could get out of the cabin, however, a husky young man in leather jacket and bulky britches, with boots that seemed laced up to his knees, crowded in, head down, avoiding all eyes.

"Uh," Kendall grunted, "my helper." He mentioned the young man's name then but the words remained so deep in his throat, I could not understand them. It sounded as if he had said: "Darned even" — a remark so bizarre that it only made the whole scene doubly nightmarish to me. I could taste the raw fear filling the back of my mouth.

The helper was a young man I had seen once or twice in Oquossoc, without being sure just where he came from. I had never known his name; later that day I learned it was Darryl Stevens. He was big, strong, and awkward, bigger than ever in his winter clothes. From his plump face and eternally sulky expression he seemed even younger than myself but he was actually three or four years older.

With all three men crowded at last into the cabin the way was clear for the McCormicks to get out the door, and after rescuing my snowshoes I followed close behind. I dared not look back lest I invite that malevolent stare once more. Although there was really not room on the trodden path for two to walk side by side, Mrs. McCormick managed to hold to Wallace's sleeve, as if she feared he might collapse into the snow or stumble blindly into the woods. They turned off toward their camp and I went on to the cookshack door where half a dozen men had clustered to await the transfer of the body. An ungainly basket of soiled wicker, a container large enough to have held Kendall himself, with a boy beside him, lay out on the snow this side of the empty pung. The teamster had managed to head the horses the other way, back toward the Oquossoc landing and he sat there hunched in his mackinaw trying to still the restless team, who tossed their heads impatiently every few seconds, making the harness rattle.

"Whoa, Billy," the man called soothingly. "Steady, Jack. Whoa there, whoa!" Steam rose from the horses' flanks and their warm breath turned white above their heads.

It struck me then that Kendall and his man must have left the basket behind when they despaired of wallowing with it all the way up through the drifted snow. Glad of a chore to keep myself out of reach, I set out to tread down a better path to where the wagon waited. I buckled on my snowshoes and followed the foot tracks, taking short plodding steps and crushing the snow down firmly. Had any creature been watching from high above I suppose he might have wondered if this was some sort of savage ritual to see me as I trod back and forth, back and forth, and back and forth again, headed nowhere, carrying nothing.

Jerry, the teamster, a dark young man with a week's growth of black beard, greeted me happily when I approached and laughed when I told him what I was about.

"Going to take some doing!" he warned me and after that he would welcome me at each return with a grin and a twist of his head. After my fifth or sixth turn, Jerry hitched the reins to the dashboard and climbed out to stamp around a little on my track, packing the snow even more firmly. I had been at this task some ten minutes before young Stevens appeared at the door of the office camp, made note of what I had done, and trudged down to fetch the basket. His feet slumped two or three times on the path I had made but most of the time he moved safely atop the snow. And when he had got hold of the basket he pulled it after him slick as a sled and manhandled it easily up the steps into the camp. The door remained open and in a minute Kendall appeared at the top of the steps. He waved one arm at the men who waited by the cookshack.

"A couple of you fellers!" he shouted. "Come give a hand."

168

There was no rush to respond to his call. The men looked at one another. More than one, I was sure, had no stomach at all for carrying a fresh corpse down over the yard. But Jim Sullivan pushed his way through and Harry Ernst came quickly after him. One of the teamsters, as if he had recalled some duty unperformed, ducked back into the entry of the cookshack, out of sight. As for me, I looked around for some retreat and could see none. If I were to abandon the track I had just finished making and hie off over unmarked snowdrifts to some other part of the campyard I would own myself either a damn fool or a coward. If I started back up the path, I might very likely run head-on into Kendall. And I could not hide in the hovel, for I had made up my mind to steer clear of Gus Lorry for the rest of my days. My only move then was to return to the lake, where I might take a stand well removed from the pung, as if I were at least ready to lend a hand. But I had no chance to draw too far away, for Jerry, as soon as I came close, called me to help hold one of the horses by the head.

"Don't suppose these critters ever *smelled* a dead man before," said Jerry. "They just might take it into their head to shy off at the wrong time. They *might* be all right. But you just can't tell. You keep old Jack's head down and give him the bit if he tries to get smart."

I was happy enough to hide myself behind the near horse and still be useful. Old Jack turned his head to fix his bulging brown eyes on my face, then stretched his neck to snuffle at my jacket. He stood in complete peace when I laid hold of the slack of the reins; I told him he was a good boy and should hold steady.

The procession from the office camp across the dooryard and out to the lake moved raggedly, and with many muffled exhortations from Eben Kendall, who had taken a handhold on the rear of the basket and tried to steer the course from there. It was impossible for men to move in a path as wide as the basket required, without some man's occasionally sinking deeper than his knees into snow beside the narrow path. The ugly basket lurched and swung and stopped. It was well, I thought, that Wallace had not waited to watch poor Tim make his final trip down to the lake, in constant danger of being dumped into the raw snow, and piped along by a stream of foul words from Kendall's lips.

When the basket had been brought at last to the path I had made ready, Kendall had the good sense to rearrange the escort, letting young Stevens go singly at the head, drawing the basket behind him like a toboggan, with Kendall keeping it straight and the others trailing to give a hand at loading the cargo into the pung. Tim weighed little enough so that two men might have hefted him easily; four men picked his weight up as if it were an order of groceries and slid the basket quickly into the pung. Jerry was there to secure the tailboard and I held my post by the horses, holding both bridles now and murmuring to the animals in the accepted way. Young Stevens climbed in over the tailboard to find his seat. Kendall stood for a time looking back up the trail for the doctor, and undoubtedly expecting to have some further word with Wallace before he made off with the body. Dr. Morton, slow as a hedgehog and carrying Tim's good suit on its hanger, picked his way along the path as if he were trying to find a crossing in a flowing brook. Kendall returned a stride or two to meet him.

"Oh, you brought the suit," he called. "Good!" But when he reached out at last to pluck the suit from the doctor's grasp he made a mildly sour face. "You could have *left* the hanger," he growled. "Well. It don't matter." He carried the suit down to the wagon and tossed it into the care of young Stevens, who simply folded it carelessly and dropped it, hanger and all, beside him on the seat.

By this time Mrs. McCormick had reappeared well up the path and Kendall, raising one large mittened hand to hold her in her place, went back to meet her. He stood talking to her in a consoling way, laying one hand on her arm from time to time, and nodding earnestly at every word she said. When he came back at last he moved, to my dismay, right in front of the horses and came face to face with me. This time he seemed to gather all his venom in his mouth and eyes and make ready to spew it at me.

"You got a goddamn big mouth!" he snarled.

I knew I had turned white. I may have formed some phrase in my throat like, "I don't know" or "That's what you think," but if I did the words never made it through my lips.

"Some day," Kendall growled, "somebody may shut it for you!"

He moved out around me then and climbed into the pung. "Let's go," he growled. Jerry shook out the reins and spoke to the team, which, glad enough to get moving again, started off at a trot — a pace more rapid than seemed quite decent, considering the errand they were on. I stood in my tracks and looked after them. Jerry had winked at me as he went by but this had not eased either my shame or my fear in the least. A sudden surge of anger, however, sent the blood to my face and neck.

"You old prick!" I called out. But whether Kendall was still close enough to have heard me, I never knew. I had not really raised my voice. Still, half blinded by anger I turned to go back to the camp and walked right into Gus Lorry, who had come down for a farewell without my noticing him. It was too late for me to go the long way around to avoid him, so I had to stand face-to-face with him on the narrow snowshoe track. But Gus would never have noticed that I was dodging him. His eyes were fixed on the vanishing pung and they actually held tears. Harry Ernst had moved over also to stand by Gus's side but there was no sorrow in his face at all.

"Well," Gus said, in a sort of stage whisper, "there he goes. The poor son of a bitch!"

"There won't be too goddamn many tears shed," Harry murmured.

Gus looked at him, his face blank with dismay, as if he had heard someone lay a curse on the dead. He shook his head and turned to watch after the pung again.

"I don't know," he said. "He had his ways. But I'll say this: He always used *me* good. *Always.* He *always* used me good."

❧ 6 ❧

THERE was no funeral for Tim McCormick, for no one was buried in the wintertime in those parts. Instead, his body was made ready, dressed in its Sunday best, and laid out for the family to view in its satin bed, and for the few friends Tim had owned in Oquossoc to say their farewells. Then the casket was sealed shut and stored away in a granite vault to await the springtime thaw. Three of the teamsters, not out of any grief or any friendship for Tim, but merely because it did not seem right to let him go wholly unmourned, rode down to Oquossoc with Wallace and his wife to murmur amen to the prayers the priest would utter and to take their leave of the body. It struck me as bitterly ironic for Wallace and Tim, who owned, as Tim himself had often confessed, about as much goddamn religion as a hedgehog, and who had often railed at the very existence of a church, that these two should have to lend themselves at last to the barren ritual that was supposed to grant them some hope of a life everlasting and a salvation they had long forsworn. Tim, the men told me, even left this earth with a cheap crucifix and rosary clutched in his frozen hand — the final indignity visited on him of course by the sister-in-law he despised. For it was only on Elizabeth's account, Wallace privately allowed, that he had acceded to all that horseshit. And this bit of barnyard liturgy, vouchsafed to Walter Hamm as he and Wallace left the Kendall Funeral Parlor together, struck me as the only phrase in the whole ceremony that Tim would have found appropriate.

I had volunteered to stay to "keep an eye on things" while the funeral party met in Oquossoc and to remain until the following day. For Wallace and his wife, perhaps to set a more seemly pace, had decided to stay overnight with the Eastwoods, an Oquossoc couple who operated a modest set of "camps" that were open to vacationers from spring to fall. The Eastwoods, according to all who knew them, were as "miserable" in every respect as the McCormicks themselves. They paid the lowest wages in the region and often found ways to chisel small corners off the final settlement. They gathered trash of all sort to treasure in a shed and never failed to scavenge the rooms the guests had left, often, the stories said, denying that anything at all had been found when a guest telephoned them to ask that a belt or a pair of sunglasses or even a necklace left on a table or a bureau be sent after them. It was the Eastwood household that shared a sauerkraut dinner with the McCormicks on Thanksgiving Day, and Sam Eastwood himself, driving a sorry horse and an old-fashioned "sleigh" such as his own father might have left to gather cobwebs in a shed, carried the McCormicks back to camp the day after Tim's remains had been tucked on their granite shelf.

It was past noon when the McCormicks returned and I was drinking tea in the cookshack, with plenty of time left to make my own way to Oquossoc and find a few nights' lodging. I did not go out to meet the sleigh but contented myself instead with watching out the tiny square window as Wallace and his wife, in their city-folk hats, which left their ears in the weather, proceeded one after the other, in silence, to the camp across the brook. Sam Eastwood, his head made round as an apple in a tight stocking cap, followed after. The gaunt horse, with a torn blanket thrown over his back and held fast by a heavy metal weight that Sam had hitched to his bridle, was left to nuzzle forlornly at a snowbank.

I still sat in the cookshack, hugging my tea, when Wallace came back in his working clothes and inquired if I wanted to step up to the office camp for a minute. I followed him dutifully and stood behind him in the camp as he worked open the padlock into the wangan room, where the books were kept. Wallace put his hat on the shelf that served as a desk and sat in the backless chair.

"I want to make it right with you for staying over," he said softly, looking directly at the wall in front of him. "What about five dollars?" "That would suit me fine," said I. Whereupon Wallace drew out the book of company orders, and laboriously, in his childish script, lettered out an order in my name for five dollars. In the box at the left on the order, he wrote "2 cds.," meaning two cords of wood, for the company advanced money only for actual production.

"Those are ee-maginary cords," said Wallace lightly, in the tone he used for telling jokes. I managed an appreciative snicker and accepted the order from his hand.

"Sam could take you down to Oquossoc if you're ready," he said. For some reason I had not looked for any such offer, and I felt only dismay. My small store of clothing was still heaped helter-skelter on the empty bunk. I had dirty underwear rolled in a ball. I had two books and three magazines I wanted to save. I glanced quickly at the disarray around me and despaired of getting everything shoved into my battered footlocker in less than half an hour.

"I got my stuff all scattered around," I murmured. "It'll take me a while."

"Sam's in no hurry. As long as the free coffee holds out."

There was nothing for it then but to begin in Wallace's presence to gather up my disgraceful clothing and to sort the bad from the good in the small heap of magazines I had collected. Eager as I had been to escape from this squalid little hideout, I found myself deliberately making much work of tucking all my goods into the footlocker, seeking more time to plan what moves I might make when I reached Oquossoc. If Sam dropped me at Haines Landing, as he was sure to, I'd have the task of toting my footlocker up the long road a mile past the Dunham place, where, chances were, Eben Kendall might be posted at one of the wide windows on the lookout for such as me. Or I could dodge him by wallowing through snow a few rods to reach the tote road. But then I would be put to the task of inventing some explanation that would persuade Sam Eastwood that it actually made sense for me to climb out of the rig before we were halfway to anywhere, and flounder off into the trackless alders. Besides I was not at all sure I could, without getting wet to the neck, manhandle my footlocker through the drifts. I had made no decision at all when Wallace announced:

"Looks like Sam's ready." I had to snap my trunk shut at last and tote it like a tray full of toys down across the dooryard. I said no good-bye to Wallace nor he to me. My only concern was deciding where I might ask Sam to set me down. Even after I had snuggled into the front seat to share a greasy blanket with Sam, I could make no choice.

"You going to Haines Landing?" I asked him finally as the poor horse moved into a rickety trot.

"Take you right to the village," said Sam. "Got to pick up my mail." Those were all the words spoken on the long ride down the tawny sled track across the ice to the landing and on past the Dunham place, where no one was about. I lifted my trunk out of the sleigh to the porch of Mackenzie's store and said my thanks to Sam, who never glanced my way.

"Glad to 'commodate you," he murmured, with no gladness whatever. He set the weight down to keep his horse from wandering, beat snow off his mittens, and shuffled into the store.

It was a gray day in which smoke tumbled out of the roof jacks and rolled down to the eaves. The small square house across the road wore a fringe of gray icicles, long as stilts. Far behind the house, on the stunted hill that had long since been scalped of its black timber, a lone man was moving through the young popple, drawing a hand sled along a trail he had beaten into the snow. No sound came from him at all.

I started off toward Jenny's at a half trot, trying to stamp life back into my drowsing feet. Sleigh tracks beneath me had become ridges of ice. An unseen bird sang two sudden notes from the fir thicket between me and the lake, and then fell abruptly silent, like a boy who had whistled in church. There was a flutter of small wings in the branches but when I looked there at the edge of the thicket, there was no bird, nor any motion anywhere. The noise of my feet, splintering ice in the deeper ruts, brought echoes from the hill.

Jenny must have seen me coming while I was still a good hundred yards away, for she came out on the brief stoop in her housedress to wait for me. Her hair hung down straight as an Indian's. She hailed me before I entered her yard:

"You fixing to stop here or you going on by?" I did not

answer but cut across the snow straight toward where she stood. "I'm frozen!" I told her.

"Well, Jesus, come in!" She opened the door at once and propelled me along with her small brown hand. "Come in and set!"

The kitchen, into which the door opened directly, was warm as midsummer. A kettle steamed lazily on the stove.

Jenny plucked my hat off my head as if I were a child, grinning right into my face as she always did. She steered me into a straight chair that she had set close to the stove.

"Get them goddamn rubbers off and get your feet right into the oven, if you like," she laughed. But I was satisfied just to snuggle close to the stove and hold out my hands to receive its blessing.

"You get fired so soon?" said Jenny, obviously meaning this for a joke. When I told her I had quit she stared at me, her lips apart, for ten full seconds.

"You have a fight?" she whispered.

"God no. I got through a little quicker than I expected though. I have this job with Kidder coming up in about a week and I was kind of laying pipe for that. Only they knew what I was up to and they . . . well one thing led to another, you know."

"Don't I ever . . ." she breathed. Losing a job in those days was a serious affair; Jenny continued to eye me solemnly, as if my brother had died.

"It's not all that terrible," I told her. "All's I need is a place to stay a few days."

"You can stay right here," said Jenny quickly, while I was still wondering how to phrase the request.

"Well, I don't know . . ." I began. But she gave me no chance to feign a proper reluctance.

"Course you will," she said flatly. "I got all *kinds* of room." As she said this, she flicked a brief glance toward the closed door of the only other room in the shack. "Where's your things?"

We dug an old hand sled out of the hovel and I used that to drag my little trunk down to Jenny's. She helped me carry it into the kitchen, where we set it by the woodbox.

"I'm going to put you in the bedroom," she said. "I made me up a little bed there in the baker." The baker was a small shed, not really attached to the house, but butted against it, and enclosing the back steps. There were rusted tools there, leaning in corners or hanging from pegs, and ragged clothes that Ernest Ryerson had left — a mackinaw the mice had been into, a pair of overalls that still held the shape of his legs. There was an old wagon wheel with broken spokes that Ernest had once brought inside to fix and had left against the wall, and boots that were worn through. Jenny had wrestled an old iron cot and mattress from somewhere and made them up, with clean muslin and musty blankets, into a bed. It was cold here as outdoors.

"I'm not going to chase you out of your bed," I told her. "I just as soon stay here. It's a hell of a lot better than I had at McCormicks."

"The hell you say," Jenny shouted. "You stay where I put you! I have to get up at night to go to the shanty anyway and I don't want to be traipsing through your bedroom." Jenny put both her strong hands on me and pushed me back into the kitchen. She opened the bedroom door, releasing a flow of chilled air, and pointed me toward it. "Now put your stuff in there and no more argument."

I lifted the trunk and set it at the foot of the wide wooden bed. My breath turned white before my face. I hopped quickly back to the kitchen and shut the door behind me.

"I'll only be a few days," I said. "But I don't feel right about taking your bedroom."

"Oh, shush!" said Jenny. "You want we should share the bed? The neighbors are going to talk anyway." I could not stop the blood from rising to my face. Jenny lifted her head to laugh her generous laugh, then she took hold of my ears and kissed me. "I'm not going to lead you astray," Jenny murmured. "Leastways not now. I got to save you for that little postmistress."

"Oh, cut it out," said I, and tried to find some chore to busy myself with. There was wood stacked in the lean-to shed outside; I buttoned up my jacket and brought in a load of maple sticks. Jenny was slicing bread to make toast, while two small chunks of ham sizzled in the spider. A Ford bounced by on the road with a tire chain beating a steady rhythm — whack! whack! whack! — signaling the bitter cold.

Jenny and I spread our small lunch out into the late afternoon, drinking four or five mugs of hot tea, the first good tea I had tasted since winter began. I have to get word to Kidder, I told myself, but the room was warm and my full stomach had rendered me too sated to go out again into the cold. So we sat there talking about the summertime while the world grew dark and Jenny had to light the big lamp. A wind had sprung up, stirring the curtains and giving a faint rattle at the window; by sunset the sky had cleared and the chilly sun flared for a time on the horizon. When the sun was gone the woods and the lake fell still. We looked out at the big thermometer on the porch post and saw that it had dropped to fifteen below.

"She'll hit bottom tonight," said Jenny.

"You'll freeze in that goddamn shed," I told her.

"Oh, I'll take me a hot brick to bed," she laughed. "You'd better take one too, 'cause you ain't going to be a damn bit better off in that drafty old bedroom. You'll have ice on the Christly chamber pot. You see." She lifted her old mackinaw off a peg and hunched into it. "Come help me knock a couple of these bricks loose."

I took my own jacket down and followed her into the shed, where for just a moment, while the kitchen warmth still clung to me, I relished the clean, sharp air. Jenny took the barn lantern down and fussed to get the wick alight, then we stepped out together in the trodden snow. The cold set us both to gasping through our teeth. There was a small pile of bricks, crested with hard snow, beside the back step. I reached to take hold of one and could not move it. Jenny laughed.

"You'd better fetch her a whack or two," she said. I kicked at the pile and merely hurt my toe. I had to go into the woodshed to find the old axe and with the blunt end of that I finally sent the bricks tumbling.

"Take some extry," said Jenny. "We'll keep some in the stove for when one gets cold." She bent down then to get hold of a brick with her free hand and just as she did the whole night seemed to split open above our heads. There was a crack like a lightning bolt, which smashed the night into a thousand pieces and sent them rocketing off the mountains far and near. At the same instant something struck the board above our heads a splintering blow. I fell instinctively to all fours and waited for the sky to collapse around me. But there was no lightning! Jenny had dropped the lantern and it now lay guttering on its side, with greasy smoke pouring out. She sat flat in the snow and stared at me with her mouth wide open.

"Jesus to Jesus!" she gasped. "That was a goddamn gun!" She scurried at once for the door, without rising from her knees, and tumbled into the shed. "Jesus Christ! Get in!" she cried. "Somebody shot at us!"

Of course it was a gun! I could hear the echoes now, racing away out over the distant woods. As I got my balance to plunge into the shed after Jenny I saw the raw splintered wound where the shot had broken the ancient plank, far above my head. For some reason I grabbed up the bricks and carried them along as if to save them from a nameless invader. The lantern lay smoking in the snow.

Jenny breathlessly held the kitchen door open just wide enough for me to slip through, then she banged it shut and slid the bolt home. I set my bricks on the floor. Jenny took hold of my free arm with both her hands and raised her contorted face to mine. She chewed at her lower lip. Her face had gone so white that the fine dark hairs on her upper lip stood out like a mustache.

"Someone tried to *kill* us!" she moaned.

"It was some crazy goddamn hunter!" I told her. Safe inside the warm kitchen, I felt no real fear at all, although I still held the metallic taste of fear in my mouth. My heart, which had seemed to stop for an instant, now beat in my ears like an alarm. Jenny pulled me close and pushed her face against my jacket. Sobs robbed her breath.

"No . . . no hunter!" she gasped. "Too dark!"

I think I was more shaken at the time by my dismay to discover that this stalwart and resourceful lady could be so instantly undone by terror than I was by the thought that someone had shot at us. Who the hell would *shoot* at us? Jenny answered my unspoken question through a sob that set her whole face to shaking so she could hardly get the word out: "K-K-Kendall!"

Kendall? I had forgotten about that bastard or had pushed him so far back in my mind that even now I found it difficult to think of him as someone who might really exist out in the dark. But now my own stomach grew small.

"He wouldn't do a crazy thing like that," I whispered. Jenny had begun to wipe her streaming cheeks on my jacket and took in several shuddering breaths.

"Hell, he wouldn't," she said, in a more controlled tone. "He thinks he owns the Christly world." She sobbed once or twice more, trying hard now to hold her chin firm. When she loosed her hold on me I moved over to the window and twisted the worn shade to peer out into the dark. Jenny grabbed me at once and tried to pull me to the floor.

"Stay away from the window!" she gasped. I allowed myself to be drawn away but I would not sit on the floor. Instead, I sat in the nearby chair, while Jenny, crouching like a child playing hide-and-seek, slid over to douse the big lamp. We sat then in semidarkness, with tiny streaks of light flickering from the cracks in the cookstove.

"You think he'll try to get in?" I whispered.

"No," Jenny growled, in a nearly normal voice. "The son of a bitch knows I got a rifle. And he knows I can use it." She raised her voice at the end, as if she were trying to reach Kendall's ear. We both sat listening for some ten minutes or more, hearing chiefly the beat of the flames in the stove and the noise of our own rapid breathing. There was a tiny rustle in the garbage pail behind me. "Mouse," Jenny whispered. We both listened to his foraging there, as he fussily explored the crannies of the crumpled butcher paper. Jenny emitted a faint snicker. "Goddamn slim pickings," she whispered. I stamped my foot and the mouse grew still. There was no sound at all from outside except the occasional gritty sweep of the wind over the dry snow.

"We'd hear him, wouldn't we?" I whispered. "If he came closer?"

Jenny answered in full conversational tone.

"He couldn't move on that crust without we'd hear him," she declared. "The bastard is gone back into his hole." There was the sound of an automobile motor in the distance, popping clearly from far across the blade of the lake.

"There he goes now," said Jenny, although it seemed to me I had heard the same sound much earlier.

"Just trying to throw a scare into us," said I.

"He managed it too, the blown-up bastard. I tell you, I like to shit." Jenny, in command of herself once more, fumbled in the cupboard for matches and set the lamp alight. She looked at me and smiled tenderly. "Wasn't you scared? I sure got a stranglehold on you!"

"I'm still scared," I said. "If it was really Kendall, I am. It could have been just some nut out there."

"Nobody around here nutty enough to fire a rifle at a lighted window. It *had* to be him." Jenny grimaced. "I suppose if he'd been fixing to kill us, he'd have aimed for the lantern, though. Hard to miss, looking into a light. Other ways than that, he couldn't have made out his front sight."

"The shot was pretty high," I said. "Did you see where it hit?" Jenny shook her head but her mind suddenly seemed far away. She stared blankly at the wall behind me, then squinted as if she were trying to squeeze something from her memory.

"You wait just a second," she murmured and moved quickly into the dark bedroom. She left the door open, so the bitter cold flowed in and made my flesh shrink. She fumbled in the dark for just an instant, then came out, her face all aglow as if she had found a hidden hundred-dollar bill.

"Can you imagine?" she cried, laughing as she spoke. "That foolish old hoptoad of a husband of mine! That was *him* out there! He never meant to kill nobody! Just his way of telling me he's still mad and ain't going to let me forget it! That warn't Kendall at all. Just foolish old knot-headed Myron White!"

This news did not strike me as anything to laugh about. I had no stomach for an outraged husband. Instinctively I got up from my chair and looked for my jacket. But Jenny, laughing in her old way, took me by both arms and pushed me back to my seat.

"Oh hell! This don't have nothing to do with you! Myron don't give a goddamn what goes on here. You rest easy. This is strictly between him and me. It's got to do with money I owe him. He *thinks* I owe him!"

"How can you be so goddamn sure?"

"The rifle!" Jenny put her laughing face right close to mine. Her breath was mild as fresh milk. "He took my goddamn rifle! He's the only one knowed I kept it hid in there. This is how he tells me. See, I got your goddamn gun. In case I hadn't noticed." Jenny kept shaking her head and laughing. "Which I hadn't! I never look at the goddamn thing from the one deer season to the next. Jesus! That foolish old critter! I wish I'd been here when he got in, I'd have put the run on him!"

Jenny sat opposite me and grinned in utter relief and joy.

"You want a drink?" she asked me abruptly. "Something good? I got a bottle put away." I shrugged and made a face to indicate I was agreeable to anything. But she had not waited for a response. She began immediately to roll back the soiled rug to uncover a small counter-sunk handle in the floor. She had to tug a little to get the trap open. It revealed only a dark hole, not more than a foot square. She reached in half the length of her arm, drew out a long green bottle, and held it up to me. It was Peter Dawson scotch, reputed to be the best in that day and in those parts, and invariably carried in from Canada.

We took our drinks in small tumblers, with extra tumblers set at each place holding cold water for a chaser. No one ever seemed to mix a drink in that day, for the whiskey was always savored, as if it were the finest wine. Yet it was often raw enough to sear a man's throat and leave him speechless. We sipped our drinks as if we were taking hot tea together.

"I'll get us some supper in a minute," said Jenny, making a promise to both of us. It was a promise she never kept. We drank our half tumblers of whiskey and she ceremoniously poured out seconds. I very soon began to experience that slight buzzing numbness around my nose and mouth that meant I was getting drunk. I remember that at one point Jenny spoke of making spaghetti and for a while something bubbled or burned on the stove. All we ever ate, however, was some bread and butter, the bread hacked off, with much laughter, from a fat loaf and the butter, too cold to spread, laid in clumsy chunks upon the bread. After that, Jenny dug up from somewhere a discarded gramophone that played old-fashioned cylinders through a big horn. The tune, thin and neverending, like the whine of a far-off saw, echoed in my mind at strange intervals for the rest of my life:

> "Rum-tum-tum-tum-tum-tiddle
> Was the tune he played upon his fiddle.
> That wonderful strain!
> Won't you play it again!"

Jenny clapped her hands to the rhythm for a moment, then began to slide around the floor in a dance step. Had I been sober I'd have stubbornly held her off when she tried to draw me to my feet to dance. As it was I rose up with great deliberation, making perfectly sure of my balance, accepted her into my embrace, and tried to follow her in the waltz, or whatever it was. Even sober I had seldom been able to dance without stumbling. Now I tripped on her feet almost at once and after only two or three sliding steps half fell against her as she backed into the wall. Without a word, she gave up dancing and slid one of her small roughened hands inside my shirt while with her other hand she pulled my head down so she could kiss me on the mouth. Even as I relished the immersion of her eager mouth in mine some sober core within my brain kept warning me that this was an unwholesome business, making strenuous love to a woman so much older. It was not how I wanted it to be, the voice was telling me.

But the urge to play the part of a *man*, an urge that it had taken two strong drinks to unleash, drove me to keep kissing her fiercely while I made a clumsy effort to get one hand on her breasts. Fondling a girl's breast was pretty nearly the ultimate in hot lovemaking in that day, at least among boys my age. Her breasts were as small as a young girl's but firm as fruit. Both nipples seemed to stand up at my touch. I heard myself gasp and I knew it was partly out of fear. I had not expected Jenny to yield up even this much of herself without at least an effort to stay my hand. Now there was no choice but to play the game right through. And I alas was not entirely sure just how the final moves were performed.

Jenny, however, knew her part well. When I had finally fixed one hand over her warm breast she pressed her hand on top of it, then kissed me quite tenderly on the cheek. "Let's go to bed," she murmured. Oh God, the far-off voice reminded me, I hadn't wanted this!

"It's too damn cold!" I don't know why I whispered. There was not a human ear within a quarter mile.

"I'll keep you warm." Jenny kissed me once more. And she began, very studiously, to undo the remaining buttons on my shirt. "Take off those goddamn boots!" she said.

I was wearing, partly unlaced, my rubber-bottom boots, standard gear for all woodsmen. How could I have hoped to dance in those? I tried to shuck them by stepping on one

187

heel with the opposite toe, but I finally had to break loose of Jenny and bend down to yank them free. I had some trouble with my balance and Jenny had to grab my arm to keep me from falling away. When I stood before her in my stockings, she laughingly undid my belt, so that my trousers would have fallen to the floor had I not caught them. Drunk as I was, I was still shy of standing naked before a woman. But Jenny had no shame. She had slipped quickly out of her own cotton shirt and wriggled free of her brassiere and undergarment in hardly a breath. She was naked now right to her deep small navel. I marveled at the whiteness of her belly and breast, next to the weathered brown of her face and neck and arms. When she kicked off her shoes they flew wildly away, one under the stove and the other against the woodbox. In one swift move, she pushed the rest of her clothes down to her ankles. All unaware I was looking now, for the first time in my whole life, at a grown girl, completely nude. Had there been a few less ounces of whiskey in my veins I know I would have flushed scarlet. Perhaps I did anyway. I was aware of a great rushing of blood to my ears and of course I looked at once to find what sexual secrets might be disclosed. But there was nothing to see but the lush triangle of ink-black hair and a trickle of black down that reached up nearly to her navel. Jenny hugged herself briefly against the chill and then moved to press herself against me, shoving her small pelvis into mine with a ferocity that, even through the insulating haze of drink, startled and unbalanced me. I could not take her in my arms, as I knew I was meant to, for then my pants would have fallen. So I put one arm around her naked waist, gripping her with at least a pretense of manly lust, and undertook to satisfy her with my kiss.

"Oh, come to *bed!*" she murmured. And then, as roughly as if she were dealing with a recalcitrant child, she tore my pants free from my grip and flung them to my ankles, leaving me standing there in my shoddy underwear, with my burgeoning penis betraying itself in a monstrous bulge. I turned away then and got myself out of my pants and stockings. But I did not quite dare shed my underwear until I had followed her into the arctic dimness of the bedroom. Swift as a mink, she slid into the big bed and let out a tiny shriek as the chilly sheets enveloped her. I had to struggle with my underwear, keeping my back turned the while, then clumsily insinuated myself, backwards, into the bed beside her. She welcomed me by trying to wrap her small wiry body around me, and fumbling to find my cock. I grabbed her hand to hold her off, for this still seemed to be, for God's sake, an outrageous intimacy. I steered my penis as best I could toward where I thought it was meant to go in and just butted it against her groin. Without warning, as if the impact had triggered a machine that controlled the entire operation, the orgasm mounted in a surge so strong and quick that it seemed to pour forth an unbroken stream before subsiding to exquisite spasms. I grabbed her wildly and held her close until the throbbing subsided against her tense body.

"Oh, you *baby!*" Jenny laughed in my ear. "You baby!"

My erection, after this, did not subside a whit. Lost finally to shame I climbed over her and suffered her to guide me properly, until I was able to thrust fiercely into the tight sleeve of her vagina.

I can smile now to recall that even as I was moaning with delight at the sensation (sweetened of course by the years of fantasy) of actually *screwing* a girl, I was flicked by a small sudden fear to realize that I was penetrating a woman who no doubt had been used by at least a dozen other men with God knows how wide a range. But the fear was drowned in a half second. I was consumed by my discovery of the indescribable. For long moments nothing seemed to exist but the sensations that flowed through my groin — the yielding moist firmness of the membranes that enfolded my cock, plus the strange friction that sent quivers of ecstasy down to my toes — a friction not quite harsh enough to hurt and yet close enough to pain to wring a noise from my throat. And then the unstoppable spasm once more that seemed to contract every muscle in my frame.

After this second ejaculation, I found myself able to perform as I had always heard (or read somewhere) that the male was meant to, pumping my member rapidly in and out to attain still another climax. This time, however, I was amazed and even somewhat frightened to feel Jenny begin to tighten her grip and strain beneath me. I suppose, like all New England boys, I had been raised to believe that girls did not enjoy sex to the degree men did, but merely permitted it. I was certainly not ready for the sounds that Jenny emitted, increasing in volume and frequency — starting with loud panting and mounting finally to a series of sharp cries like "Uh! Uh! Uh!" toneless as the barking of a seal. They took me so aback that I stopped my thrusting instantly, afraid I had done her an injury. But Jenny dug her nails into me. "Don't stop!" she gasped. "Don't stop!" So I resumed, beginning myself to feel the irresistible tingle that signaled the approaching climax. And Jenny, clutching my neck in a final spasm, let go with a scream that surely must have sounded out in the lake. "Oh Jesus!" she yelled. "Oh Jesus Jesus! Oh my GOD!" Through all her outcries I kept pounding away, feeling my own climax hastening and when it came I too cried aloud. Then we both lay limp for a few moments while echoing spasms stirred me. "Oh, my loving Jesus," Jenny breathed. She gripped me tightly then and hugged me with all her strength. "That was *wonderful!*"

I had no strength to move or speak. I seemed to be sinking forever into a warm deep well of bliss, still coupled with Jenny but unaware of it and unaware even that I had arms and legs.

I awoke with a freezing cold nose. My eyes opened rather stickily and I looked up at a strange ceiling that seemed lightly streaked with smoke. Paint had come free in a curl just above me. How, I wondered idly, did I get here? And where is it? I pulled part of the musty bedclothes — a quilt with the white stuffing breaking loose at the edge — up to comfort my face. My breath rose up like steam from a kettle. If I had not needed so suddenly to find a place to piss, I might have sunk back into the far-off world that had enfolded me. I stirred uncomfortably. I was damned if I would use the pot! But where? Little by little I reoriented myself and little by little the doings of the night before came back to me. I was naked. I reached down and discovered that dried semen had caked on my groin and thigh. A flush of shame burned my ears. Where the hell was Jenny? If I could get up and get hold of my clothes and get away before . . . But the sound of a stove lid being replaced told me that Jenny was up and busy in the kitchen. For some reason I had a notion that she would be angered and ashamed. Well, I was ashamed too. But now I *had* to get outdoors before I burst. I saw my forlorn underwear hanging from the bedpost, limp as a mop, and I grabbed it frantically and held it against me as I slid out of bed. I reeled for just a moment as if the drunkeness still clung. But after a few false thrusts I was able to get my legs and arms enclosed. Then I yanked the quilt off the bed, wrapped it around my head and shoulders, and fled for the kitchen, barefooted and shoeless, shrouded in my tattered quilt, like a homeless Indian. Jenny turned from the stove to greet me with a happy and generous smile. The kitchen was hot, and sweet with the smell of fresh coffee. Something hissed in an iron spider on the stove.

"Well!" said Jenny. "You get tired of resting?"

"I have to take a leak! Quick!"

"Here," said Jenny. She gathered up my boots from beneath the stove and dropped them by my feet. I jammed my feet inside, letting the laces dangle, and shuffled for the back door. "Anywhere by the woodshed!" Jenny called after me.

The zero air seized me by the legs as I pushed my way out into the sun, and along the well-trod path to the woodshed. I moved just far enough to be out of sight of the road and relieved myself upon the frozen snow, sighing as if my whole body were being deflated. A violent shiver rattled my frame. I had to struggle to keep the quilt on my shoulders and I kept shivering as I emptied myself, absently eyeing the scarred chopping block with the taped-up axe stuck in it, the rusty bucket hung atilt on a nail, with the pulp hook beside it, the coil of haywire tossed on top of the tumbled pile of graying "junk" wood, the ancient carbine leaning against the neatly piled stovewood, the worn blue-handled bucksaw, the five-gallon kerosene can with the wooden plug in its spout and the name *Peabody* painted on it in fading black letters. There had never been anyone named Peabody hereabouts since I first came.

The moment I returned to the kitchen Jenny held both hands out.

"Now before you do another damn thing you got to get yourself into this tub and get a hot bath! You must be covered with *gunk!*"

I could only stand and gape at her. I could feel the mortification setting fire to my face. Jesus! To sit naked in a tub before her eyes!

As if she had read my mind — not a difficult thing to have done — she backed away and showed me where she had placed the little oval galvanized tub. "You can squat right in there and I'll keep my back turned every *minute*. You got soap there and a big towel on the chair. Now don't argue! I ain't going to peek at you."

It did seem like a happy idea, for I had not properly bathed since I had hiked to the storehouse, days before Tim died.

"You'll be getting to smell like one of them old lumberjacks!" Jenny laughed, keeping pace with my thought.

"I got to get some clean underwear." I hurried into the cold bedroom and fished a new union suit out of my little footlocker. When I came back, Jenny kissed me, holding my head between her hands.

"I hope I didn't spoil you for that little postmistress," she whispered.

"Why don't you cut out that crap?"said I. I had a growing urge now to take myself far off, away from Jenny and this dingy little shack, and from any reminder of what had happened to me last night. And I had no appetite for conversation that might seem to ratify any permanent relationship between me and Jenny. I stepped free of my boots and, still shivering a little from the outdoor cold, began to unhitch my underwear.

"Who owns that old rifle out there?" I asked.

It took Jenny a second or two to reply.

"What old rifle?" Her voice was half a whisper.

"There in the shed. That carbine."

So as not to invite Jenny to look at me I kept my face turned away. Now there was a longer silence.

"Oh, my Jesus!" she gasped finally. "That's right! I left that son of a bitch out there!" Her voice broke now. "Oh, my Holy Jesus! It *was* Kendall. Oh, Christ! Listen. I've got to get away from here! That wasn't Myron! It was Kendall! Oh, where the Jesus will I *go!*"

I was almost out of my underwear but I clutched it against me and turned to look at her. She had her lower lip held tight in her teeth and her nostrils were wide with fear.

"He's not going to get you in broad daylight," I told her. With the clear day and the climbing sun I felt no fear at all. There were cars moving and someone up the road was chopping wood. The train would pull in before the sun was much higher and the whole village would come alive. "Anyway, if he'd really wanted to *kill* you, for God's sake . . . Oh, come on. He just wanted to throw a scare into you."

Jenny expelled a long sigh.

"I suppose so. Just the same, I ain't going to stay around where he can find me so easy. He might take a notion to scare me again. He does too good of a job of scaring folks."

"Where can you go, anyway?"

Jenny looked me steadily in the eye as if she were considering whether she should say what she had in her mind. But she merely shook her head briefly and turned away.

"You go ahead and get your carcass clean, I'll think of something. First off, I'll fetch that goddamn gun."

She put her jacket around her shoulders and slid out the door, careful not to let in too wide a blast of air. I quickly rid myself of my clothes and squatted in the tub, gasping as the hot water stung me. I rose up halfway to escape, then squatted quickly again when Jenny came back. She was able to laugh now at the sudden way I had sought to conceal my nakedness.

"I'm not going to peek!" she cried. She carried the gun into the bedroom, admitting a slice of air that ran across my back like an icy hand.

"Jesus Christ!" I yelled. "That's cold!" Jenny squirmed

back through a stingy opening and shut the door quickly behind her. She kept her eyes studiously averted.

"You find the soap?"

"Uh-huh." The soap was right beside me on the floor and I began to lather myself generously, splashing about as I did so until puddles grew on the bare floor all around. The heat from the cookstove blessed my upper body and I settled deep into the embrace of the sinfully hot soapy water, as snug as if I had been in my childhood bed.

"Want I should rinse you?" With my shame now properly sheltered within the deep tub and the gray water, Jenny was watching me like an indulgent aunt.

"Stay the hell away," I told her, but she came over anyway and made as if to empty the steaming tea kettle over my cringing head. She laughed at my yelp and then held the towel for me. Still shy of exposing myself I would not stand up until she had turned her head away. I dried myself on the big rough towel, while the heat from the cookstove beat on my back.

Jenny had made toast and bacon for breakfast and I sat down, with my hair still damp and tousled, and helped her eat it. Each time I looked at her, as she sat cuddling her teacup, she was eyeing me in that mildly speculative way, obviously making herself ready to ask me a question. I believe I had half a notion of what she was thinking but I could dig up no remark right then to aim her toward another subject. She set her cup down on the table and leaned toward me.

"When are you fixing to go up to Kidder's?"

"Well, I don't know. Whenever he's ready. Pretty soon anyway." I avoided her eyes. This was exactly what I feared she would say.

"I think I might just take a notion to go with you."

The dismay must certainly have shown in my face, even

though I pretended to be absorbed in putting extra butter on my toast.

"It can't be *that* bad," Jenny murmured. She looked at me with lips protruded, as if she might start to cry.

"Oh, no! No! It's just that . . . well, I don't know what Kidder . . . I'd have to ask him." Ask Kidder! I flinched at the very prospect. But Jenny just snickered.

"Oh, *that* old critter! You don't need to ask him! Just *tell* him. Tell him that I'm coming to see he minds his manners. Why, I knowed old Kidder since he was driving a twitch horse for Tague. Yes, and before that!"

"I think I should ask him all the same. You can't just barge in . . ."

Jenny's face had grown very solemn indeed.

"Don't you fret. He ain't going to turn me out when I'm in trouble. I'm dead serious. If they's one man in this whole county I'd want around me when I'm in a fix like this, it's Jim Kidder. No goddamn Kendall's going to scare him. But, by Jesus, he'll put the fear of God into Kendall. Nobody around here wants to tangle with Kidder. You ever see him handle a gun? Son of a bitch! He can drive nails with a forty-five! From fifty yards! I seen him! Old Kendall get funny with him, by Jesus, Kidder'll pick the buttons off his vest!"

Jenny's eyes were burning so and her face had grown so grim that I had to laugh.

"Jesus, Jenny! You're not going to start a gunfight?"

"No, I ain't. But if that old shithead of a Kendall starts one, by the Jesus, Jim Kidder'll finish it!"

"Well, I still think he was just trying to throw a scare into us."

"All the same, I'm going to sleep a hell of a lot easier with Jim Kidder handy." Suddenly she put her strong little hand out and laid it on mine. She broke into a fond grin. "Not that I didn't sleep easy with you!"

I could feel the fiery blush mounting to my hair. Jenny laughed out loud and kissed me.

"Oh, don't feel bad," she whispered. "Don't feel bad. I ain't going to lay any claim to you. Now let's get our stuff together and go surprise old Jim."

"I don't know where I'm supposed to go."

"Then call him!" Jenny gave a shove to my arm. "Call him!"

I swallowed the rest of my tea and got myself into my coat.

"Comb your hair first," said Jenny.

❦ 7 ❦

THE weather had turned almost warm by the time we met Jim Kidder, and there was sweat under my shirt. I hadn't told Jim I was bringing Jenny with me, although I said on the phone that "we" would meet him at the time he set. He just hadn't asked who it was that made us we.

When we came to Bill O'Brien's place, where Jim was to wait for us, there was no Jim in sight. The pung was pulled up in front of O'Brien's shack and the sturdy horse was tossing his head about as he tried to snuff up the last bits of grain in his nose bag. There were half a dozen cartons of assorted groceries in the pung and a long ice saw and a chisel. We had no more than reached the muddy path that led into O'Brien's door when Jim came out of the shack and hailed us, as if we were still a few rods away.

"Well, Jesus!" he cried. "Jenny! You running around with this critter now?"

"Just trying to keep him out of harm's way," said Jenny. She had planted herself near the pung and came no farther. With her red stocking cap pulled down over her ears and the black hair protruding all around like fringe on a curtain, with her eyes narrowed and her mouth solemn, she seemed like a stranger, ready to turn away if she was not instantly welcome.

"Jenny wants to come with us," I said.

Jenny's eyes widened.

"If you got room," she murmured.

"We'll make room!" Jim declared. "Even if we have to sleep Bob in the shanty!" Jenny smiled at this, but not in her usual manner, with her lips open and her face alight. Her mouth just turned up for an instant and she made no sound. To see her so tamed this way, as if fright had taken all the spirit out of her, filled me with a sudden tenderness that made me squirm. I felt an urge to grab the slack of her jacket and pull her close. Yet all the way over from her cabin, dragging the little hand sled with our private wangan on it, I had been prickly with worry that she might crowd too near, even hang to my arm as if we now belonged together.

Bill O'Brien appeared in the open door of the shack and beamed on us both with the indiscriminate friendliness of a drunk. I saw Jenny suck in her lips at Bill's appearance.

"Come in, Jenny, for Christ's sake!" Bill's voice was as throaty as a raven's. "Come in by the fire!"

Jenny merely shook her head and I hastened to explain that we had to get going, although I could think of no reason why. There were at least four hours of daylight ahead of us. Bill paid no notice, for he probably had not really wanted us to come in at all.

"You going to cook for this outfit? Christ, Jenny, you can cook for me! I'll keep you a hell of a lot warmer than Kidder will!"

Jenny's only response was a baleful stare. Kidder had taken the nose bag off the horse and was ready to go. I piled our duffle in the pung and set the small sled on top. There was room for all three of us on the seat. The horse shook his head to readjust the bit and Kidder slapped the reins on the horse's back.

"C'mon!" he shouted. "Get up in that collar!" And the horse, his head nodding as if he were towing a ton, started off with such spirit that Jenny and I had to grab the seat to keep from toppling. Kidder laughed.

"Old Billy knows when he's headed home!" he said. "Hang tight." The snow in the roadway, packed and polished by sled runners, gleamed like glass. We sped into the deepening brush and over the gentle hump of the corduroy bridge, where the woods grew deep and where the clop of hoofs, the tuneless jangle of harness bells, and the random thumping of the sleds died without an echo.

"I just as soon that that old son of a bitch O'Brien not know I was coming up here," said Jenny. "He's got a big mouth."

"Oh, hell," said Jim. "He's just setting out to tie one on. Give him a couple of days and he won't know if he seen you day after tomorrow or six weeks ago next Sunday. Or if he seen you at all. I'm the one you got to watch out for. I still don't know what the hell this is all about. Not that you ain't welcome. *Anytime.* But what you leaving your happy home for? Old Myron gunning for you?"

"Not Myron. Kendall."

Kidder turned to stare her in the face.

"*Kendall?* That old pisshead? What'd he do to you?"

"Took a shot at us," I said. "Anyway, somebody did last night while we were out back of Jenny's place."

The reins lay slack in Kidder's hands.

"Son of a bitch!" he breathed. "You sure it was Kendall? Not some jacker?"

"Wasn't any light," said Jenny, "and nobody else's got any call to shoot at either of us. Far as I know."

Kidder looked grimly at me and nodded as if we were sharing a vow.

"That old prick shows his face up here at the club and I'll ventilate his ass for him. Right, Bob?"

"You're the man to do it," I said.

"What's he after you for? He running short of customers?"

Jenny shook her head glumly. We looked into each other's eyes and saw the same doubt there. I held my peace until Kidder turned to examine my face.

"He thinks we know something, I guess."

"You wouldn't have to be too goddamn smart to know more than that dumb bastard. You find out where he hides his liquor? Or what?"

Jenny and I looked once more into each other's eyes. Jim shook out the reins and spoke mildly to the horse: "C'mon Billy!" Then he turned to watch us expectantly.

"I *always* knowed where he kept his liquor," Jenny muttered. She grimaced at me. "Might as well say what it is."

I turned to face Kidder.

"He has an idea we know about that killing. About Jack Dunham."

Kidder gave an almost silent whistle.

"You think *he* killed Dunham?"

"I don't know a damn thing about it."

"By Jesus, *I* know!" said Jenny. "It just stands to reason. It had to be either him or that Christly Bailey. The two of them been doing up Kitty Dunham's chores for a year now. Every time that poor goddamn old Jack's been off scaling somewhere. He must have come home once when he wasn't supposed to."

Kidder laughed.

"You could send a man to jail on *that* kind of evidence and you'll have half the county behind bars! Right, Bob?"

I felt a hot blush climb my neck but Kidder did not seem to notice. He gave full attention to the horse, who was laboring up along the rise now with plumes of steam chugging from each nostril.

"Who else is going to shoot the poor critter?" said Jenny, her voice vibrant now as she came out of her gloom. "Talk about Frenchmen, for God's sake! What are they going to kill him for? His money? Kitty never let him keep more than a quarter in his pocket."

"Who knows what kind of a rig she was running?" Kidder had raised his voice to his customary shout. "Maybe she was taking on a few Frenchmen between times."

"Oh shit!" Jenny exclaimed and fell silent, as if that were answer enough.

The hill we climbed was a long one, traveled but little, although the snow had been trod down enough to make the going easy. We had turned off the main tote road to the storehouse and I had not noticed where.

"Where we headed?" I asked.

"You never been up to the club?"

"What club? I never knew there *was* a club."

"Winnebago Club. You been up here before, Jenny?"

"Once years ago. My sister did chamber work here and she brought me up for a day. I loved it."

"There ain't much to love up here this time of year. Except me and Bob."

Jenny laughed as cheerfully as ever.

"Well, I'll try to make do!"

The cold had begun to take hold now, here where the sun could not find us. We snuggled close to each other. The black growth on this side of the lake had all been cleared years ago, leaving the lean hardwood, barren in winter, but growing densely enough in spots to make thickets of alders that roofed the road like a canopy, shaped by the snows of twenty winters. The dogged horse kept nodding and twisting his head as alder branches, thick as fingers, whipped his ears. Out from the alders finally, where young birch and poplar hedged the road, two chickadees appeared to keep us company, skipping from twig to twig quicker than fish, with a sudden purring of wings. "Dee-dee-dee," they told us as we passed, speaking, I suppose, of good things that lay ahead.

It was a long, long climb to whatever hideaway the club had chosen. After an hour, Kidder stopped the horse and helped us tuck the heavy blanket around our shoulders and over our legs.

The final stretch was a level one through a thicket of second-growth fir not much higher than our heads, where the air smelled suddenly like Christmas. The snow alongside the road had been deepening all the while and now it hid most of the undergrowth. Then the woods fell abruptly away on our left and we looked down across an expanse of untracked snow that dipped in gentle billows to the pond. Pencils of brush, bare of buds, poked through the snow to mark the water's edge. The pond was two miles long, reaching straightaway from us and widening as it went, a long pure stretch of white, blemished only meagerly by dark spots where the wind had brushed the snow away, and embraced on both sides and at the upper end by the solemn evergreens, soldier-straight, that seemed, from our distance, to grow thick as moss.

The club was a scattering of log camps, now all nearly eaves-deep in snow, reaching out randomly from a two-story lodge that commanded the slope. A squat chimney rising from the lodge roof wore a bonnet of snow atop a set of short planks that kept the weather out. Beyond the lodge, where the road was leading us, a low roof lay, almost even with the snow, the shingles nearly bare. A shiny roof jack spilled a lazy flow of smoke down across the wet shingles.

Kidder spoke softly to the horse.

"Whoa, boy." The horse settled to a stop, shaking his harness bells. "We got company."

"Oh, Jesus!" Jenny's voice was just a breath. Kidder chuckled.

"Well, it ain't Kendall. I'll guarantee that."

"Don't be too goddamn sure. You don't *know* that son of a bitch."

"Well, I know he couldn't get his fat carcass up that last rise without he had a crew to haul him."

"What if he had a hoss?"

"No hoss been up this road since it snowed. And how the hell would he know where you was headed?"

"I guess you're right. But I'd be a lot easier in my mind if you'd go on ahead."

"Oh, for Christ's sake, Jenny! If it *is* Kendall, which it ain't, all the better. I'll drive the son of a bitch right out into the snow and you can go pick up his frozen carcass in the morning. What part the foxes ain't got."

Jenny laughed and Kidder clucked the horse into motion. The road dropped sharply as it passed the lodge and the little horse had to hustle to keep the pung from catching up. Someone had dug a long trench in the snow from the back door of the lodge to the cabin behind. The road turned uphill again as it neared the cabin and Kidder pulled up at the turn. We all held our places on the seat until the cabin door opened and a short woman, dressed in a man's wool shirt and heavy trousers, walked out on the covered porch.

"Hey *Jim!*" she cried.

"Well, Jesus Christ!" Jim shouted. "The things you see when you ain't got your gun! You taking charge here now? What you done with Taylor?"

"Oh, he's on a toot. You know how he gets. I had to clear

out. Then King said he thought you could use someone over here to do for you for a few days."

Kidder glanced at us, offering a quick grin, as if we had a joke to share.

"King's always looking out for my welfare, ain't he?" he told the lady. "Well, I'm going to be up to my ears in help. You know these critters? This is Bob Smith, been scaling for McCormick. You must have heard of him."

"Can't say as I have," said the lady, grinning up at me. Light glinting off her heavy glasses made her stare seem blank. "Hello! I'm Lilla Foster." I pulled off my mitten to shake her hand. Her grip was muscular as a man's.

"The bathing beauty at the end is Jenny White. Jenny's one of them Magalloway skunks."

Lilla reached far over to meet Jenny's mitten.

"Ain't he terrible? I declare, I don't know where that skunk business started. I always say some of the best people in the whole area come from up around Magalloway. Ones *I've* known, anyway."

"It's my husband comes from over that way," said Jenny cheerfully. "I was raised right down here in Rangeley."

"There now! You see!" Lilla fetched Kidder a wallop on the knee with her bare hand. "You just can't believe a word this one tells you! I don't know why I listen to him!"

"Well, the way I see it, you marry a guy, you take on his nationality. That right, Bob?"

"That's the law," said I.

Lilla waved one hand at me in pretended disgust.

"Oh, you're two of a kind! One lies, the other swears to it! Jenny, what're we going to do with this pair."

"Hitch them up in the hovel with Billy!"

"Now that's a *grand* idea. I got plenty of halter rope too!"

"Better get going, then," said Kidder. He shook the reins out over Billy's back. "C'mon, BILLY!" And the little horse, who had been shifting his hooves impatiently and blubbering his lips, made off again with a start that nearly unseated us.

It was too close to dark now to do anything but make ready for supper. Jim Kidder surrendered the "winter camp" to the two women, then he and I, with me on a pair of borrowed snowshoes, set out to tread down a solid path in the snow to a small board camp on the rise beyond. This was a dank little den, made nearly dim as a cave by the snow piled deep against the rear window. There were two iron beds here and a tight wooden chest fortified with zinc where mattresses and bedding were secured from squirrels and mice. Jim set a fire going in the small box stove, then we retreated to the main camp to share the heat of the kitchen. Jenny was moving deftly and silently about to get the long table set for four, while Lilla, fussing by the stove, talked on and on, in the high-pitched happy tone of someone who has just returned from a journey. Jenny greeted us with a small grimace and a slight tilt of her head toward Lilla, who had acknowledged our entrance by waving her hand toward the table while hastening to finish the sentence she had addressed to Jenny. She picked a saucepan off the stove and shook it to rearrange the contents, took hardly a second to replenish her breath, then addressed us all, while she lifted a cover to peer into the spider, where some sort of meat was sizzling.

"I was just telling Jenny," she called out, "about the time we was smelting that spring, the four of us, up Cupsuptic. You remember that? Didn't it *rain?*"

Kidder grinned in her direction.

"I remember they was a lot of liquor consumed."

"Oh, Lord, *wasn't* there? But still not like the time my cousin Eben come along. Was you there then? When he fell in, boots and all?"

"Eben who? Not Kendall, for Christ's sake?"

Lilla, holding a large spoon in one hand, turned to face us.

"That's right. The mortician down to Oquossoc. He's an own cousin to me. Didn't you know that?"

"No. And if I was you I wouldn't brag about it. Not in this company."

Lilla looked from one to the other of us, frowning slightly in disbelief. For some reason she centered her attention on me.

"I know they's some folks don't like him. For business reasons." The way she said this made it almost into a question.

"I don't even know him," said I, speaking part of the truth.

"I never cared for him," said Jenny glumly.

"Well, I declare!" Lilla now bent her gaze upon Jenny. "I know he *can* be difficult. Specially when he's been drinking. But when you get to know him! Really, he's the most *comical* man. Seems strange too with the work he's in. I think *that* turns a lot of folks off." She looked almost beseechingly at us now, obviously hoping that this was something we could all agree on.

"Well, *I* never did business with him," Kidder declared. "And I ain't looking forward to the day I do!" This gave everyone a chance to contribute a small chuckle.

"Well," said Lilla brightly, "can't say as I blame you."

When eating started, Lilla was almost the only one who spoke at all. Jenny seemed to have little appetite, and when our glances would meet across the table, hers seemed full of foreboding. Lilla spoke mostly of the "old days," some ten years before, when she and Kidder had worked, if not together, at least for the same employer at the storehouse— the great supply depot for all Brown Company subcontractors who cut pulpwood or moved it millward throughout that area that embraced the Kennebago lakes and the Rangeleys, with their myriad tributaries, dams, and outlets. There was only one other major operator on that side of the "height of land" — the International Paper Company, or "I.P.," all of whose wood was moved down the same river, the Androscoggin, that flowed past the Brown Company mill at Berlin, New Hampshire, and the I.P. mill at Livermore Falls.

When we had all finished our pie, Lilla protested she needed no assistance in doing up the pots and dishes, yet happily accepted our six extra hands to dry and put away all the tools of cooking and eating. Because Jim Kidder knew better than Lilla did just which cabinets and shelves held what dishes and pots, he took over the storage job and soon had the whole kitchen in order.

"If you fellers felt like playing cribbage or something," said Lilla, "I'm sure they's a board here somewhere with cards. Seems to me I seen it when I was straightening up."

Jenny spoke for the first time since the meal began.

"I'll go up and make those beds up for these two," she declared and seemed not to hear when Lilla tried to protest. She hunched herself into her coat and stepped into the weather. I knew that she meant for us to come along so she could let go all that had been simmering inside her since supper began but it seemed cruel for all of us to stamp off together.

"I'm going to hit the hay pretty damn soon," said Kidder. "A man's got to be pretty well rested up before he tackles Bob here at cribbage."

I had never in my life played cribbage in Kidder's presence and had never played it at all before coming to Maine. Kidder winked at me and I tried to make a sound like a laugh.

"Well, I'd better go up and get my junk out of Jenny's way," said I. "I spread it all over the camp."

"I'll take the top bunk!" Kidder shouted and filled the cabin with his laugh. I ducked out quickly and left Kidder to think up his own excuse.

Jenny had not even touched the bedding when I came in. She was sitting on one of the two wooden chairs. The camp seemed colder than outdoors and Jenny's breath rose in a thin steam from her slightly open mouth.

"I got to get the hell out of here!" she declared. "That bitch will have her goddamn cousin up here before you know it."

"How can she do that? He doesn't know where she is."

"Well, he'll damn soon know where I am, once Lady Big Mouth gets to a telephone. I'm going to get Jim Kidder to carry me out of here tomorrow."

"Oh, come on, Jenny. Why would she want to tell him that? Anyway, it must be five miles to a phone."

"Just the same . . ." Jenny muttered.

I set out to liven up the fire in the tiny box stove. Kidder came in while I was still stripping bark off birch sticks to hurry the flame. He had just closed the door behind him when Jenny spoke up.

"You've got to carry me out of here tomorrow!"

Kidder squinted at her.

"Now hold your hosses! Nobody's going out of here tomorrow or the next day either. We got work to do. When I take the pung out next I'll take old Lilla with me, and if you get lonesome for her, why you can just bring down that old talking machine from the shed and crank her up. We still got a few cylinders somewhere."

"I'm not kidding!" Jenny cried. "That old bitch will have Kendall up here as soon as she finds a telephone!"

"What'll she tell him? That she met somebody that don't like him? Anyway, her place is to hell and gone below Arnold Pond. That's thirty miles as the crow flies from Kendall's place. More than that. More like forty. Fifty miles by the road. She going to call him long distance to give him the glad tidings that a couple of herring chokers up at Massachusetts Gore think he's full of shit?"

"She's bound to go tell him something, with her big mouth. Or spread it around the county anyway."

Kidder took Jenny's hand in his.

"Now listen to your uncle Jim. There's no place in the state of Maine where you're better off than here. If by any wild chance that old pisshead finds his way up here, I'll run his ass out before he can scratch it. Besides, when I do get back to the storehouse I'm going to get hold of Stapleton and tell him about your being shot at. Believe me, old Stape will give that old bastard religion if nobody else will."

"Oh, Jesus! Don't do that!" Jenny actually squirmed with dismay. "You'll only make it worse! You don't *know* that goddamn Kendall . . ."

"The hell I don't! I knowed him when he was walloping pots at Bald Mountain! He always talked big. And he always had shit in his blood. One hard look from old Stape and he'll piss himself!"

Jenny had taken a tight hold of Jim's big hand in both of hers.

"Well, I wish you wouldn't, Jim, please! I don't want to get him any more stirred up. Just because he's scared of Stape don't mean he's scared of me."

Jim pulled Jenny to her feet and got both his arms around her. She pressed her head to his chest and held him tight.

"Anyway, here's where you're safest. Bob and I will keep old Eben off you. Right, Bob?"

"I'll do my damndest!"

I was mildly annoyed to discover that the sight of Jim and Jenny in tight embrace sent a splinter of jealousy through my stomach that almost made me wince. What the hell did *I* care?

Jenny laughed very lightly and freed herself from Jim's grasp.

"I know. I know," she said. "But you know that old bastard put one hell of a scare into me. You ask Bob. That shot came *that* close!"

"Maybe it wasn't even Kendall," said Jim. "Could have been that other joker, whatever his name is."

"Bailey? Oh, he wouldn't have nerve enough to throw a snowball. No. It *had* to be Kendall. Who else could it be?"

"Couldn't be anybody *I* know," Jim patted her cheek. "Everybody loves you, Jenny."

"Well, by Jesus, I don't love everybody."

Jim grinned at her.

"Christ, no! You ain't had time!"

Jenny gasped.

"Kidder!" She drew her arm back and fetched Jim a wallop on the chest with her open hand. Jim ducked away and raised both arms, pretending to guard himself from future blows. His laugh filled the room like the roar of a fire. It still seemed to be echoing when the rapping came on the door. Jenny stiffened as if she had been stabbed, and looked around, terrified. The door opened instantly to admit Lilla, loosely arrayed in a mackinaw and wearing no hat. She was smiling as if she had shared the joke.

"Land's sake!" she cried. "What's going on up here? Such carryings on! I thought I'd better get up here and join the fun!"

Jenny's face grew sullen, while Kidder and I looked quickly at each other, not knowing who was meant to speak.

"Jenny's taken to slapping us around," said Jim. "We just don't move fast enough to suit her."

Lilla made a grimace.

"Oh, come on now. Jenny just ain't that *kind*. I can tell a

211

nice person when I see her. You let them talk about you that way, Jenny?"

"They can talk any way they like," said Jenny, her cheerful face still glum.

Lilla became immediately solicitous.

"You fellows have hurt Jenny's feelings now, with all that rough talk. I declare, I think when you fellows come out of the woods they'd ought to wash your mouths right out with yellow soap. Now why don't you two get right out of here and let me and Jenny red this place up . . ."

"No! No! No!" Jenny cried, offering Lilla her quick, bright smile. "I can do better without no help! Goes better that way!"

Bewildered by Jenny's sudden change of mood, Lilla backed off slightly. She shrugged and made a slight grimace.

"Well, all right. I know how it is sometimes when you have your own way of doing things. Seems like the other one's always . . . I know I . . ." Lilla's voice trailed off and she forced a prim smile. "Well, I'll just get out of the way here and go make up Jenny's bed." She turned toward the door, then looked back at Jenny. "I'm putting you in the little room right next to mine. They's a cot in there they probably use for company. Only thing is, if you get up at night, you have to come through my room."

She widened her eyes at Jenny and seemed to await a response. But Jenny had taken the hem of a pillow into her mouth as she prepared to slip it into a case and she could only grunt her agreement.

"I sleep sound!" Lilla assured us all, then she bundled herself out the door. Her boots squeaked on the trodden snow.

Jenny, the pillow neatly stuffed into its case, smacked it hard to make it plump. She spoke darkly, in a voice not meant to reach beyond the door.

"I can stand about one night with that mouthy pig. Then I'll cut my throat. If I don't cut hers first."

"Oh, hell," said Kidder, "she'll be out of here in a day or two. I ain't finished all my chores back at the storehouse and I'll lug her back with me when I go. Less'n you got so fond of each other by then, you can't stand to lose her."

"Two days," said Jenny, "and I'll be chewing up the bedclothes."

"Maybe one day," said Kidder.

"That a promise?"

Kidder stared at her for a moment.

"Okay. I was hoping to get Bob started on the ice. But I suppose I might as well get this blight off your life. Make it tomorrow forenoon. If the weather holds."

But the weather did not hold. The wind rose that night, howling off the border mountains and setting the black growth all around us to bending and moaning like monsters in a dream. The snow began before dawn and pelted like blown sand against our window. When we opened the door in the half-light to attend to our toilets, the blizzard struck us in the face and scattered snow across the floor. It was bitter cold. By the time we were snuggled into all our clothes there was a foot of new snow on the path to the main camp. Nearly blinded in the wind, we took temporary refuge on the porch and found the red mercury in the big thermometer outside the door had shrunk almost completely into the ball at the bottom. It was fifty-five below zero, the lowest temperature I had ever known.

But it was warm in the kitchen, where Lilla fussed at the stove and Jenny was methodically plunking tableware into place along the table.

"How'd you like this for weather?" Lilla cried, turning her flushed face toward us as we stamped in together.

213

"Close the door!" Jenny implored us, although I already had it half closed behind me.

Jenny and Lilla, in strict accordance with New England usage, poured coffee and set out platters of food while the men sat down to be waited on. There was bacon enough for a boy scout troop, a full dozen doughnuts, home fried potatoes, and, of all things, hot rolled oats to begin. When everything lay in front of us, Lilla sat herself down and beamed over her handiwork.

"Well," she gasped, "you fellers think you can make out a breakfast?"

Kidder stood up suddenly.

"That reminds me! I got to go feed that Christly hoss!"

All three of us laughed.

"Bring him in!" Jenny shouted. "They's plenty to go around."

I was well into my cereal when Jim came back. He had taken time only to put on his wool hat and mittens on the way out and he hugged himself and shivered as he shut the door.

"Jesus! A man could lose an ear out there! Don't you girls go picking berries or anything without you got your longies on!"

"Bet that hovel's cold," said Lilla.

"It ain't so bad. Old Bill has spread shit enough around to steam it up."

Lilla put one hand to her mouth.

"Oh, Jim! Not at the *table!*"

Jim grinned at me.

"I wasn't planning no such thing."

Jenny spluttered into her cereal and put her napkin to her lips. Lilla turned red and shook her head in mock despair.

To my own amazement, I was able to tuck away the rolled oats, devour two helpings of bacon and fried potatoes and finish two doughnuts and two cups of coffee before I stopped to stretch. I ate one more doughnut after I carried my dishes to the sink.

The wind relented when breakfast was over but snow still fell, straight as rain and hard as hail, covering all our tracks to more than a foot. The cold closed down tight as a fist.

"I ain't going to break no trail today and freeze my balls off," said Kidder.

We stood together in the porch of the main camp, where an embankment of snow shut us off from the world. A deep growl arose from the lake, as new ice, forming deep in the lake's belly, tried to make room for itself. A crack like a rifle shot sounded from somewhere behind the camp and I turned to look at Kidder.

"What the hell?" said I.

"One of them big trees freezing. They'll snap like that when they freeze to the heart."

"Well, I say let's get in where it's warm."

We kicked the new snow aside to make a path to our camp and set a fire going there. Even with the flame roaring up the pipe, every breath we gave turned into white vapor.

"They keep a cribbage board here somewhere," said Kidder. "You play that game?"

I confessed I did and we played cribbage all morning, with Jenny sitting close to watch us. "I had all I need of that lady," she told us. "She like to talk both ears off my head." But when Kidder broke off the game to go tend the horse, Jenny returned to the kitchen to lend a hand with dinner.

In the night the weather cleared and the heavens filled with frozen stars. Jenny had come to sit with us until bedtime and when we had begun to nod she moved reluctantly to the door.

"I hope to God you can get her out of here tomorrow," she murmured.

"Oh, I guess Billy could make it if it'll just moderate a little. He's seen more snow than this. I'm of two minds about leaving you here with Bob, though. No telling what you'll get into when nobody's watching."

"Oh, shush!" Without a backward glance, Jenny stepped out into the snow, pulling the heavy door tight behind her.

"To tell you the truth," Kidder said, in a hushed voice, "I wouldn't fight *too* hard if she tried to climb into my bed. She's something extra, that gal. She can do anything a man can do. And a few things he can't."

"I know," I said, although I really did not know any such thing. I fumbled about in my mind for some sufficiently casual way to toss in the fact that she *had* been in my bed, or I in hers, but I could find no phrase that fit.

In the morning, when Kidder undertook to break a trail of some sort down to the lake, Jenny was right there to take part, even grabbing hold of the ice chisel before I could take it from Kidder. Once on the ice, we shed our borrowed snowshoes, for the wind had swept the lake surface almost free, except for windrows here and there where the snow came over our ankles. Kidder, who seemed to have picked out the exact site for the operation, walked directly to a half-clear spot some twenty feet from shore and began to shovel away what snow was left until he had cleared an area about the size of somebody's front porch.

"See if you can rassle that chisel away from that young lady," he called to me, "and I'll show you where to start."

Jenny pretended a reluctance to give up the chisel but I plucked it from her mittened hands without effort. There was still wind enough here to lift snow off the ice in swirls and spit it into our faces. The cold still brought water to our eyes and kept us turning from time to time to take the wind on our backs.

"Make me a hole here about eighteen inches across. Just keep a-chopping till you find the water. Don't be surprised if you have to go down two feet or more." Kidder turned to us and tried to work his stiff lips into a smile. "When you got it open, you can drop Jenny down through to take the bottom end of the saw."

Jenny, standing with her shoulders hunched and her chin tucked tight into her coat collar, responded with a snort.

"If'n you don't get that other critter out of here real soon, I'll jump in there myself."

Kidder handed Jenny the shovel and set off to recover his snowshoes.

"You coming up to say good-bye?"

Jenny did not even look his way.

"I told her good-bye already."

"Well, just keep that hole free of ice and spell Bob when he begins to buckle."

"Like hell," said I. I had already started to break the ice at my feet and Jenny shoveled the chips away. We stopped when the jangle of harness sounded on the slope above us. We could make out only the muffled heads of Lilla and Kidder above the snowbanks. Lilla lifted one hand and her shrill voice sounded like the scream of a jay:

"You two take care now! Don't stay out in the cold too long!"

Jenny lifted one hand in salute and I yelled my good-bye. Then Kidder shouted to the horse and the whole array seemed to sink into the snow. We could hear the harness bells for two or three minutes, jingling crisply in the stillness.

I found pleasure in chopping off great jagged chunks of ice with each jab of the chisel, while splinters of ice flew right and left. Jenny, impatient to be moving, kept stamping her feet to keep the warmth in them. After about ten minutes, when I had made a hole as wide as a barrel and about three inches deep, Jenny set her shovel down.

"I'm going up to dress my feet better," she panted. "My toes is about ready to drop off."

She hustled into her snowshoes and made her way painfully up the steep trail we had left. I kept on chopping with the joy a boy takes sometimes in smashing large pieces of crockery. I lost track of the time that passed before Jenny appeared at the top of the snowbank and half slid her way down.

"They's someone coming!" she gasped. And there was real terror in her tone. I looked up to see if I could sight something in the road where Kidder and Lilla had gone.

"No!" Jenny cried. "Up the lake! Way up the end!"

"I don't see a damn thing," I told her. There was nothing there but white expanse, bordered by the darkness under the snow-laden spruce.

"Wait! Wait! You'll see him. He's in that cove." Indeed, just as she spoke I saw the figure — small, straight, and black — obviously moving our way.

"Let's get out of here! Come on! Quick!"

"Well, Jesus! Let's see who it is first."

"I don't want to see who it is! What if it's Kendall?"

"Coming from up there? What the hell would Kendall be doing up there? Coming by way of Canada?"

"Just the same, I don't want to make no target out here. You come too. Come up inside! I got my rifle!"

"You toted that rifle all this way?"

"Course I did! Now will you come on! Christ's sake! He'll be here in twenty minutes!"

Although I earnestly desired to stand and wait until I could make out the man's face, I went over to join Jenny. I shoved my feet into the snowshoe clips and just shuffled along behind her.

"Oh, come on!" she beseeched me, almost weeping. So I hurried to stay with her, looking back only once to watch the approaching figure. His legs seemed unbelievably long. Jenny held the kitchen door for me and shoved me inside.

"That's not Kendall," I told her when she had shut the door. "Not with legs like that. And whoever it is, he'll come right up here anyway. He can't miss our trail."

"Well, at least we'll be *inside*, where we can see him first."

Jenny went immediately into her little room and brought back her carbine. She sat at the table and in less than a second had a shell in the chamber. I had to laugh at the scene she created, like something from a William S. Hart movie.

"You going to *shoot* the guy?"

Jenny looked slightly abashed.

"Well, I ain't going to be caught like last time."

We sat a long while in silence, listening to the steady ping of melting icicles dripping on the metal cover of the gas tank outside the back window. I felt no apprehension at all. Actually I sensed a mild relief that I was not going to be left to cope alone with Jenny, who might, for all I knew, assume that I was going to share her bed. And like most kids my age, I always relished the arrival of company — or the ringing of a telephone, for that matter.

It seemed like an hour before we heard the man on the porch of the main camp, his snowshoes clattering on the thin carpet of snow that lay there. Then, almost instantly, his form darkened our window. Jenny laughed aloud and set her rifle promptly on a chair.

"It's Bill Jones, for God's sake! Old Bill! Toting a load of booze most likely!"

I knew Bill Jones well, for he had been a game warden in our area and I had even once helped him find a guiding job at the hotel. He never had a chance to knock at the door, for Jenny opened it in his face and we both stood there to hail him. Bill was an extraordinarily tall man, perhaps six-foot-four, and so thin he seemed like a caricature. His legs looked long as pickpoles, and not much meatier. And he had a nose that seemed as long and as sharp as the beak of a loon. He was panting from his hike and kept nodding to us in greeting. There were two streams gathering on his upper lip and tears ran down both cheeks, so he seemed to have been weeping from pain, or from fear. But he was calm as if he had simply answered a call for dinner. He looked from one to the other of us, frowning slightly.

"What the hell brings you folks away out here? You got a job with the club, Jenny?"

"We're just giving Kidder a hand with the ice," said Jenny quickly. "Come in for God's sake. Get out of the cold."

Bill slipped the pack easily off his back and set it beside the door. It took him nearly a minute to unbuckle his snowshoes and he studied me all the while, as if he had some news to impart.

"I thought you was with McCormick," he said finally.

"I got through."

Bill nodded and stuck his snowshoes in the drift beside the steps.

"Good thing," he muttered.

Jenny hurried to put out cup and saucer and doughnuts and had it all on the table before she asked him, "How about some good hot tea?"

"Go good," said Bill. He was unwinding himself painfully from his clothing. His knee-length britches were plastered with snow and the stockings, rolled just below the knee, were iced all around the cuff. He pushed his stocking cap back but left it on his head when he sat down to blow on his tea.

"Never figured to see you so far from home," he told Jenny. "You hiding from the law?"

Jenny studied him for several seconds before she answered.

"Well, I'm hiding from *somebody*. Some joker took a shot at me."

Bill pulled his head far back and widened his eyes at her.

"Took a *shot* at you? On purpose? Who the hell done that?"

"Well, I don't know for sure. But I got a pretty good idea."

"I didn't know as you had an enemy in the world. What the hell you been up to since I seen you last?"

"Well, I got *one* enemy I know of. And I'm staying clear of him."

"You tell Stapleton?"

Jenny shook her head.

"Didn't want to take the time. Bob and me, we just gathered ourselves up and flew!"

Bill turned at me with his eyes wider still.

"You in this *too*, for God's sake?" He squinted at Jenny. "This ain't got nothing to do with *Myron*, now?"

Jenny laughed out loud.

"Oh, hell, no! You think he gives a goddamn what I do? No, it sure as hell warn't Myron. He don't have no rifle, for one thing. It was a big lard ass who thinks Bob and me been spreading tales about him. Which we haven't."

Bill was studiously dipping his doughnut into his tea and consuming it in large chunks. He talked into his teacup.

"Sounds like Eben Kendall to me."

Jenny's mouth dropped open.

"I never said that!"

"It's just that the dee-scription fits. Did you ever think he's just out to drum up a little trade?" Bill laughed, mostly to himself.

"I don't think it's so goddamn funny," said Jenny sharply.

"No. It ain't funny. But I wouldn't put it too far past that old pisshead. I never had a hell of a lot of use for him, you know. Son of a bitch tried to get me fired once after I took him for bait fishing in Quimby Pond. Bastard had fifteen trout on him, five over his limit. Son of a bitch! He offered me a five-dollar bill. Can you believe that? 'This is just between you and me,' he said. 'No, sir,' I says. 'It's between you and me and the folks at Augusty.' Well, Jesus! If you ever seen a man turn white!" Bill stopped to savor more doughnut. He was clearly well wound up to follow his yarn clear to the end.

"Now wait," said Jenny, getting to her feet. "Bob and me has got to get on with that ice job. You want more tea?"

Bill shook his head. Being suddenly untracked this way seemed to leave him wordless.

"You can rest up in that little room there," said Jenny. "Maybe you'd ought to dry your clothes by the stove."

"I'm all right," said Bill. "You think there'd be a bed for me for a night?"

"Sure thing. Take Kidder's bed, up there with Bob. He won't be back for a day or two."

"You sure it's all right? I could pay something. A little anyway."

"Oh, shush," said Jenny. "We're glad of the company. You could lend us a hand with the ice when you're rested."

Bill, his mouth full of doughnut, nodded vigorously.

"Change my stockings," he said. "Be right there."

When Jenny and I returned to our job the hole seemed shallow indeed. Jenny took a few shovelsful of ice chips out of it and it deepened a little but there was surely a good eighteen inches to go to reach water. I began to chop more lustily and I quickly set several large slices adrift. Then my chisel seemed to hit solid rock and I let out a sudden groan. Jenny laughed.

"You're through the snow ice," she said. "Now you start on the *ice!*"

"Two feet of *that*, for Christ's sake?" I took a firm grip on my chisel and drove it with all my strength into the ice. This time I cracked out a chunk twice the width of my hand.

Chopping through this iron mantle was not nearly so zestful as the job of destroying the snow ice. But I pounded away and after a time began to knock larger and larger sections free, where Jenny could deal with them. But the hole sank with painful slowness and my muscles protested the steady jolting, so that when the appearance of Bill Jones on the ice provided a good excuse, I stopped and watched him shed his snowshoes and move out to lend us a hand.

"Where you going to stow it?" Bill asked.

Jenny and I looked at each other in mild bewilderment. The icehouse. Was it even visible in the snow? Finally Jenny lifted a mittened hand to indicate the place — a small building with an unpainted face, roof deeply mantled in snow, with a gaping doorway where the snow had piled almost halfway to the top.

"You're going to have some job getting that sawdust out," said Bill. "She must be froze a foot or more. You can't stow the ice there until you clear that out."

"Oh, Jesus!" Jenny looked at me and bit her lip in a sort of apology. "Well, Kidder wanted this hole cut anyway. Just to get the job started."

We all walked over to the icehouse then to appraise the size of the job that awaited us. Bill and I both wallowed up to the doorway and looked into the dark. There was indeed a small mountain of sawdust there, lightly crested with snow. Bill climbed in and fetched it a solid whack with his boot. It gave not an inch.

"You'll have to take the axe to that to get it started," he said. "But once you get the top broke off, she'll come in chunks."

"I vote we leave that to Kidder," said Jenny. "All he wanted from us was to get the hole open."

"I can make the first cut," said Bill. "If you got the saw."

"We'll have to wait until Bob gets the hole open. But the saw's up by the hovel."

"All right," said Bill. "I'll take over here and you can go up and keep that fire going."

Jenny seemed happy enough to surrender her shovel. She beat her mittened hands together and trotted back to get her snowshoes.

"I'll fetch the saw when you need it!" she called. "Just give me a holler."

Bill took a spell with the chisel after a few minutes and he was soon breaking off chunks the size of small boulders, which I quickly shoveled up and sent sluicing across the ice all around us. When I took the chisel back, the hole was well over a foot in depth and the ice broke in neat jagged sections from the sides. I whacked away, breathlessly, eager to be first to reach the water. But it was several more minutes before the chisel plunged suddenly almost the whole length of the handle and water began to surge up as if a hydrant had been opened. I chopped away to widen the opening, while the water leveled off and silently climbed the sides of the hole.

"There, by Jesus," said Bill. "You done it!" He lifted out some floating chunks of ice in his shovel and sent them skittering. We grinned at each other. Tears ran down both sides of Bill's long nose.

"You get something in your eye?"

Bill took one greasy mitten off and wiped at the tears with a finger so grimy it must have gone unwashed for a week.

"No," said he. "My eyes always water. Been that way all

224

my life." He gave a deep sniff and shook his head. "What's going on between you and Jenny?"

I felt the shameful blood scald my face and neck.

"Nothing!" I declared. And at the next moment I wished I might have given a more manly reply — a wink and smile, perhaps, and some sort of wisecrack. But guilt had overcome me in a breath.

Jenny brought the long one-handled saw down when we shouted and she stood by as Bill began the first cut. He drew the big saw slowly up and down with a rocking motion, edging it inch by inch into the new ice, while I used the chisel to widen the well of water we had made. Five or six tiny black creatures about the size of ladybugs darted on random pathways through the water like something seen under a microscope, the first outdoor life to appear since the bitter cold. The cold had moderated since morning and as I chopped away at the side of the hole, sending the water splattering, sweat trickled from under my hat.

With the first cut finished, Bill began a new one, following a path slightly on the bias, so that the two cuts began to converge. I spelled him then for a while, rejoicing to see the big saw eat so relentlessly into the hard ice. When I had drawn my cut out to the same length as the first one, I pulled the saw out and fished up my bandanna to wipe sweat from my face.

"Hot work," said Bill. He was fussily trimming the edges of the hole and seemed to be studying the antics of the tiny bugs. He pursed his lips suddenly and sent a dark brown stream of saliva out onto the ice. It was the first time he had spit in an hour or more, and he must have held that quid in his mouth since he came down.

"Might as well snake out a few hunks. Won't do no harm if they do freeze." He brought his chisel over and deftly chopped a connection between the two cuts. After only a half dozen whacks a whole strip of ice bobbed up suddenly, like a toy boat in a bathtub. Bill used the chisel to edge the wide end out into the hole.

"They's sure to be a pair of tongs up there in the ice-house," he said. "Fetch it, will you, and we'll yank a few of these out." He turned to look up at the camps. "Must be getting close to dinnertime."

I had to wallow through the drift to get into the dark ice-house, where I saw the tongs at once, hanging from a spike.

Bill had patiently sawed off a two-foot chunk of ice and after a few faulty efforts to hook the tongs on cleanly I yanked the piece out of the hole and set it sideways on the ice. It was very nearly square. Bill gave it a nudge with his foot to position it properly, then sawed it neatly in two.

"Haul easier," he said, before I had a chance to question him. By the time Jenny called us to dinner we had six cakes laid out on the ice, ranged about as if someone had toppled a tower of blocks.

Left to herself, Jenny had cooked up one hell of a dinner. Kidder had brought two chickens in and Jenny had stewed one of them with dumplings. Whatever else she had found to toss into the pot had filled the small room with fragrance. Bill put on a show of sniffing the air, shaking his head all the while in open approval. Jenny set the small washbowl in the sink and brought the kettle from the stove to pour out water for us to wash in, but Bill walked straight to the cookstove, took the lifter from the side where Jenny had set it, picked one cover off the stove, and spit another long brown stream into the fire, along with a small cud that looked like something a dog had swallowed.

Jenny screamed.

"Jesus Christ, Bill! Not when I'm cooking! My Jesus!"

The fire hissed briefly as Bill set the cover back.

"Can't eat my dinner with a cud in my mouth," he muttered, without even turning to look at Jenny. But I looked at her and for two or three seconds shared her horror. Bill meanwhile shed his jacket — a short mackinaw that might have been red once but now was all smudged, so it looked as if it had been used to wipe off an engine.

"Where you got me?" said Bill, putting a tentative hand on one of the chairs.

"Ain't you going to *wash?*"

"Don't need to," said Bill. "Ain't done nothing to get dirty." He lifted his hands slightly as if to exhibit their purity. "Had my gloves on all the while."

Jenny, noting as I did that both Bill's hands were as grimy as a blacksmith's, simply let her mouth fall open and shook her head at me in speechless dismay. Bill pulled the chair out and sat down while I went to the sink to wash.

Hungry as I was, and earnestly as I applied myself to my own plate, I could not resist frequent glances to watch Bill as he now and then fished up a chicken bone from his plate and, with gravy trickling between his fingers, almost daintily nibbled it free of meat. I heard Jenny's tiny gasp as Bill licked the stray gravy from his hand and I dropped my own glance quickly to my plate so as not to meet her eye.

Like most men in the woods Bill had no time to waste in talk while there was food in front of him. But when he had consumed two full helpings of chicken and chased the last bit of dumpling about with his fork until he could impale it, he leaned back in his chair and put both hands on his belly.

"By Jesus, Jenny. I got to give you top marks for cookin'. How'd that goddamn fool ever let you get away from him?"

"He didn't have nothing to say about it." Jenny spoke in a half whisper, clearly indicating that she was not going to dwell on that topic. She gathered up our plates in both hands and brought back two large cuts of apple pie, which she had concocted out of canned apples. It was still hot

227

from the oven. Bill finished most of his share in three bites and then toyed for a minute or two with the bits of crust that had eluded him. When he had done away with those, he reached into his pocket, drew out a black plug of tobacco, and worried off a small portion with his stained front teeth.

"Now listen, Bill," Jenny cried. "If you got to spit, spit into the heater. Not into my cookstove, for *God's* sake."

"Don't have to spit," said Bill.

Looking closely now at Bill's soiled shirt, whose cuffs wore a ring of grease that must have been gathering all winter, at his wrists that were no cleaner, I winced to imagine how Kidder would react when he came back to find his bed had snuggled a body as foul as that one must be. I knew the stockings were fresh. But from the aroma that Bill gave forth now that we sat in this close space together, I was certain his underwear was the same he had climbed into at the end of the summer.

Jenny probably had not thought anything of that, but she was clearly eager to get Bill out-of-doors, where the winds might freshen him.

"What you going to do about that booze?" she said. "You going to leave that lay there till Kidder comes in? He might not be too pleased to find that there."

"Booze? What booze?"

"Didn't you bring down a load of booze? You got a full pack."

"Christ, no!" Bill exclaimed. "No booze, for Christ's sake! Stockings! Wool stockings! I can make as much on a gross of them wool stockings as I can on fifty quarts of scotch!"

Bill turned toward me and dealt me a wink. His eyes were streaming.

"A goddamn sight easier toting too! Right?"

I agreed that he was right.

228

8

BEFORE Kidder came back, the thaw had really taken hold, and all the roofs in the club ran water. Snowshoeing was nearly impossible, for the snow clung like mud to the snowshoes and turned the webbing to something like wet string. The trail we had made to the pond, from tramping back and forth, had turned icy and now a thin stream of snow water ran its whole length. Snow dropped in sodden clumps from the trees, and the forest seemed shrouded in fog.

Bill Jones, who had planned to take off for Oquossoc after breakfast, stood for a long time on the porch and appraised the prospect. It would be heavy going on snowshoes, he decided, even on the road, which was a route Bill preferred not to follow. His usual trail took him down an abandoned tote road, then along the frozen stream to Little Kennebago, where there was no one at this season to mark his passage. But he was eager to be on his way before Kidder came, for fear Jim might bring a passenger or two who would make too much of Bill's presence. The lawmen — and there were only two in the area — probably would not give a damn about stockings, there being no statute forbidding their simple possession. But Bill preferred not to answer questions. He *had* run a lot of booze over the border in his day.

I stood beside Bill on the porch and savored the almost springlike air.

"They's something about a fog," said Bill, "that cuts the snow." I did not dispute this bit of lore, although it seemed backwards to me.

"I'd really ought to get out of here," he murmured. "But shit!"

He did not need to complete his thought, for he had expressed it to me four or five times since daybreak.

"Be a good day to get that sawdust out of the icehouse," said Bill. His tone was halfhearted, almost as if he wished I would veto the notion. But eager as I was to see Bill on his way before Kidder came and viewed the desecration of his bunk, I could think of no sensible reason for postponing the sawdust job.

"I'm ready," said I. So together we put in the whole forenoon breaking open the frozen sawdust pile and heaving ragged chunks of it, like giant portions of cork, out through the door. The core of the pile ran loose as gravel and we took shovels to clear it. When the job was done, it was close to dinnertime, so we trudged back to the camp, carrying our coats, which we had had to shed in the warm air. Jenny opened the door to greet us.

"I was just about to give you a holler. You got that icehouse all neatened up?"

"Should have left it outside to begin with," said Bill. "No need to handle it twice."

"Well, these sports like things tidy." Jenny offered us her generous smile. "Don't matter to them how much trouble they give the help."

Bill nodded.

230

"All millionaires here, ain't they?"

"Many times over, some of them."

"I don't have much truck with them folks," said Bill solemnly. "I guided them some, back along. Jesus. You'd a thought they owned the Earth. Which I suppose they pretty near do."

"Oh, they's some nice ones. Not too many. But I knowed a few when I was here before. Whyn't you shuck those boots out here, so's you won't nasty up my floor? You got sawdust all over them."

We both hastened to obey and came into the kitchen in stocking feet, like sleepy children.

After dinner, we cut more ice. There was a thin film of water on the lake surface and we did a lot of sliding about. Bill stopped once and raised one hand for silence while I lent an ear to what the breeze might carry.

"Thought I might have heard them harness bells."

I listened and heard nothing but a tree squeak nearby. Bill shook his head.

"Don't suppose they'd come anyway in this sticky going. He'll wait for the freeze."

Whatever he waited for, Kidder did not show up until hours after Bill, in the frosty dusk of early morning, started for the tote road, one cheek full of tobacco, and his leather pack high on his shoulders. The wind had come up, bringing the cold weather back, and the crusty snow made good going. Jenny and I meanwhile, having discovered, when its ridge of snow had melted, the long sweep, or crane, that had been built to hoist ice blocks off the lake and into the icehouse, undertook to drag a few blocks over, hitch the tongs to the arm of the sweep, and swing the blocks into the icehouse. A bag of rocks tied to the other end of the sweep made just counterweight enough so that the blocks could be hoisted with little effort.

We had not filled the first layer when we heard the harness bells. We abandoned the job at once to hurry to the camp dooryard to greet Kidder. There were two figures on the pung with him, so swathed in clothing they could have been either sex. But the crumpled sombrero that Stapleton wore identified him at once. Jenny caught her breath.

"Oh, Jesus," she whispered. "I asked him *not* to . . ." She drew back a trifle as if she might seek a place to hide. But of course that would have been senseless now. We both stood in plain sight as the horse drew near.

"I brung company!" Kidder shouted to us. At this, the slight figure in the middle raised his hand and called out: "Hello there!"

It was young Curtis, whom Kidder had brought with him to McCormick's. For some reason, his sudden reappearance made me uneasy.

I suppose I cringed at the prospect of sustaining a conversation about "college," which Curtis had suggested when we parted last. I had not stayed in college long enough even to memorize the names of the buildings where the different classes were held. But I waved a greeting anyway and managed a smile when he hopped down from the wagon to grasp my mittened hand. His clothing — red plaid mackinaw and matching knee-pants — looked as if he were wearing it for the first time. His feet broke through the crust and he sank over his boot tops in the snow beside the road. He still held my hand as he floundered to find some footing.

"Nice to see you again," he told me, giving me his practiced smile. His face, bright red from cold, shone like a schoolboy's. Unable to dredge up any response formal enough to match that greeting, I muttered something like "All right," and as usual felt the embarrassment surge to my neck and cheeks. But Curtis had turned at once to Jenny, had ripped off his wool hat, and was telling her happily that he was Gordon Curtis. Jenny found his dressed-up manners comical and laughed in his face indulgently as if he were a four-year-old successfully imitating his elders. But she turned her head away almost at once to keep track of Stapleton.

Stape rode right on with Kidder as far as the hovel and gave Jim a hand in unhitching the horse. With nothing more to say to each other, we all three stood there and watched the two men wrestle the harness free.

"He sure knows his stuff," Curtis breathed, in that half-hushed tone some men use to indicate awe, or reverence, or wonder. It was as if we had just witnessed some startling athletic feat. Having harnessed and unharnessed horses by the dozen, I could not keep from snickering. Curtis turned to me with an expression so pained that I almost told him I was sorry. Instead I simply murmured, "Sure does," and hoped it did not sound as if I was making fun of him.

Neither Jenny nor I exchanged a word with Stapleton until we had all gathered in the warm kitchen, had shed our clothes, and were sitting at the table, where Jenny distributed cups of coffee. Stape had removed his worn jacket but he still wore his ancient hat pushed back in a sort of acknowledgment of his being indoors. With its bent brim pointed skyward, the hat gave him a slightly startled look, as if his hair were standing on end. But his expression was inscrutable as ever. He looked from Jenny to me. He took a sip of coffee, held it in his mouth for a moment before swallowing, and then spoke directly to Jenny:

"You think Kendall took a shot at you."

It was not a question but a statement. Jenny, bent over her own cup, just shook her head.

"I ain't accusing *nobody*."

"You *think* he did though."

Jenny now looked Stapleton in the eye.

"Well, by God, somebody did. You can come down to my place and I'll show you."

Stapleton nodded four or five times.

"I know. I been there." He fumbled in the breast pocket of his shirt. "I got the slug." He held out his hand to reveal a small splayed and ragged chunk of lead, like a broken filling from a giant tooth. "Thutty-thutty," he said. "Only Kendall never fired it."

Jenny could only gape. Stapleton was holding the slug between thumb and forefinger, studying it as if it were a gem. He spoke without looking at Jenny.

"You said it happened Tuesday?"

"That's right. Long about suppertime."

"Kendall was at lodge meeting Tuesday night."

"That what he told you?"

"I never asked him. I seen him there. He had supper at the Pines."

Jenny's face had gone white and she looked as if she might start to cry.

"Then who . . ." Jenny's voice broke and she had to swallow hard. "Who else? . . . Who'd want to *kill* me?"

Stapleton sighed deeply and shook his head.

"Beats me. But that's what I aim to find out."

"Well, Jesus!" Jenny cried. "What am I supposed to *do?*" Her voice rose sharply and burst out in a sob. Stapleton reached over and put one hand over hers.

"You just stay put, that's all. Ain't nothing going to happen to you while Kidder's on the job." He looked intently at Jim as he said this and Jim leaned forward too and reached one hand out toward Jenny.

"You're goddamn tooting," he said. "Nobody gets within a mile of this place without we give him a going over."

"Somebody come down over the lake in the last day or

234

two," said Stapleton. Undoubtedly he had marked the faint trace of Bill Jones's snowshoe track across the ice.

"Old Bill Jones," said Jenny in her normal tone. Then she added hurriedly, "He'd been trapping up by the line."

Stapleton chuckled.

"Running booze, more likely."

Jenny shrugged.

"He may have been running something. But it warn't booze."

Stapleton patted Jenny's hand and grinned at her.

"Don't matter a goddamn anyway," he said.

Gordon Curtis all this while had been sitting with his mouth slightly open, staring at Jenny as if she had sprouted long ears. He looked from Stapleton to Kidder.

"Did someone really try to *kill* her?"

Stapleton stared back at Curtis for several seconds, then shook his head.

"Maybe just trying to put a scare into her."

"But *why*, for God's sake?" Gordon seemed to address his question to the whole company.

"It's a long story," said Kidder.

"Somebody thinks I know something, I guess," Jenny said, in a voice just above a whisper.

Curtis looked from one to the other — apparently amazed that there was not more turmoil.

"Well, my God!" he exclaimed. "Shouldn't I get *Dad* into this? He ought to know what's going on."

"No need for that," said Stapleton flatly. "This ain't got nothing to do with the company."

"I know. But my God. Dad would know what to do maybe. I mean, he has lots of drag down in Augusta and around. He could get some people up here to investigate."

We looked at each other silently, seeming to exchange the

same unspoken thought. Finally Stapleton gave his little dry chuckle.

"That's my job, son. And I ain't run out of gas yet."

Curtis blushed.

"I know. But after all . . . if there was more guys . . . I mean . . . Well, if they knew *Dad* was interested . . ."

His eyes wet, his generous mouth tortured with embarrassment, and his frame, with his bulky mackinaw off, revealed in all its frailty, Gordon seemed more than ever like a child. Jenny, her own spirit reviving, looked at him so tenderly I thought she might take him in her arms.

"Well, it's real nice of you all the same," she said.

Curtis recovered himself at once. Assuming his college boy solemnity, he nodded several times and spoke in a tone about half an octave lower than customary.

"You just say the word, and I know Dad will be *glad* to help. He'll be at the storehouse in a few days and I could get him up here in two shakes."

Kidder, out of Gordon's line of sight, made a face of mock dismay. I knew what Jim would say about the prospect of any such visitation if we were by ourselves. I had a hard time keeping my face solemn. But Jenny managed to look grateful and Stapleton's face remained expressionless. He studiously set his cup square into the saucer, laid the spoon at its side, and carried all to the sink. Everyone in the room was watching him. For the first time I noticed that he wore a holstered revolver on his left hip, covered by his jacket.

"Believe I'll just poke around for a spell," he said. He settled his gaze on me and seemed just about to smile. "How's the snowshoeing?"

I realized of course that he needed no advice from me.

"Good," said I, in standard Maine style. Stapleton gravely accepted my appraisal, pulled his hat down so it covered the tips of his ears, and moved quietly out the door.

236

"Quite a rig, that guy," said Kidder.

"I don't ever want to be on the other side from him," said Jenny.

In the three hours remaining before dinner, Kidder and Curtis and I divided up the ice job, prying the blocks loose from the surface where they had become frozen and towing them on the hand sled over to be swung into the icehouse. Kidder awarded Curtis the task of hooking the cakes to the sweep by means of the ice tongs and Curtis, once he had learned the trick of driving the tongs into the ice so they would hold, took to the task with an almost demonic frenzy, floundering about on the snow as he guided each swinging cake home, then clumping back in his gleaming boots to hook on to the next cake before I could beat him to the job. When we went back to cutting, Curtis tried just once to work the big saw through the new ice but could not seem to keep the blade from buckling, so he turned to towing the hand sled, almost angrily refusing any help when the sled did not budge as easily as he expected it to, then panting along with his hat pushed back, his face agleam with sweat, as if he feared the ice might turn to water before he got it safely snuggled away.

"Whoa, boy!" Kidder told him after his second trip. "We got to make this job last."

Curtis, who was wide-eyed with a sort of frenzy, stopped then to laugh and blow out a long breath. From somewhere in his gleaming jacket he pulled out a neat white handkerchief, shook it out of its perfect folds until it was nearly the size of a flag, and daintily dabbed his forehead and face.

Jenny appeared then carrying two empty buckets.

"Less'n you fellers want to wash up in snow somebody better fetch me up some water. I'm scraping bottom up there."

Curtis, who stood closest to her, grabbed both buckets

from Jenny's hands and set out to dip both at once into the hole at our feet. They bobbed about a little at first, so that Curtis had to set one bucket down and use both hands on the job of filling them. Kidder meanwhile stood grinning at his side, ready to handle the second bucket. But Curtis would have none of this. He pushed Kidder's hand away and lifted up a brimming bucket in each hand. At his first step, the buckets slopped over and doused his pant legs.

"Oh, shit!" he gasped. He looked quickly at Jenny. "Oh, I'm sorry," he told her.

"You better let me take one of those," Jenny laughed, "or you'll get to cussing."

Curtis shook his head vigorously.

"No! No! No! I can make it."

The water had apparently begun to seep into his boots, for he lifted his feet one after the other and shook them.

"Come on," said I. "You'll get soaked."

Curtis resisted only a little as I worked his hand free of one of the handles. I dumped about a pint of water on the ice and started off, with Curtis, who quickly emptied the top off his own bucket, plodding along behind.

"Goddamn it, Jenny," Kidder shouted. "You're taking my whole crew. What the hell are you up to? You ain't going to lead them astray?"

"I ain't going to lead them anywhere. They got two more trips to make. At least. Before I get the tank filled in that range."

Jenny, in her oversize boots, followed along at our heels and showed us where to empty the buckets into the stove. Four more buckets filled the dark well just about to the top.

"Might as well get these kettles filled too," said Jenny, "if you fellers ain't wore out. Then I could use a bucket of cold, to thin out the high wine."

"High wine?" Curtis looked at both of us in turn, squinting his eyes in puzzlement.

"That's what they call alcohol from Canada," I told him. "It's a hundred and eighty proof. That's ninety percent alcohol. They cut it with all kinds of stuff. Ginger ale, Moxie, Coca-Cola."

"Jeepers. I learn something everyday. I know in school they have a lot of this Belgian alcohol. The guys mix it with lemonade and stuff. High wine. They really mix it with water?"

Jenny smiled her tenderest smile and patted Curtis gently on the cheek. I was dismayed again to feel a sharp twinge that almost brought a twist to my lips. Was I *jealous*, for Christ's sake?

"Not up here they don't," said Jenny. "Not in my kitchen. Anybody runs a can of that stuff in here and they got Stapleton to deal with. All's he'll do to you is ship you down to Farmington for a nice long rest in that little brick boarding house."

Curtis, who had flushed at the touch of her fingers on his cheek, had pulled away slightly. He squinted at me again.

"I don't get it."

"The county jail. Stapleton's the law up here."

"I know. But I didn't know that Stapleton . . . Dad thinks he's a kind of Toonerville cop."

Jenny looked deeply shocked.

"Stapleton? Don't kid yourself. I'll stack him against any of these federal gumshoes we see up here. Tough? He'd go hand to hand with a barnful of bobcats."

"Well, I didn't know. I suppose Dad never saw him in action or anything. It's just that . . . Well, you know, coming from the big city. Well. Cops are different there. . . ."

Jenny laughed.

"Only time they ever got Stapleton into the city he like to choke. Couldn't find a spring!"

When we were toting our last two buckets up to the kitchen, Curtis still had water on his mind.

"You *drink* this stuff? Isn't there a well or something?"

"Spring is frozen. This water won't kill you."

"I don't know. All those animals and fish crapping in it. And bugs!"

"No different from some of those big reservoirs."

"Maybe not. All the same. I think I'll give it a miss."

"It's been boiled over and over by the time you get it in your coffee."

"That's true. But I'm not going to chance it for drinking." He pondered his decision the whole way up to the door. Then he said firmly that he guessed he would take his drinking water from the steaming kettle and cool it in the snow.

"Or drink the snow," said I.

"That's an idea."

"Only you might find a rabbit turd."

Curtis snorted. "You got some great ideas. I'll stay with the tea kettle."

When we returned to the ice, Stapleton had appeared, moving swiftly toward us down the white lake, almost as if he were skating. All three of us stood and watched.

"Look at him *come!*" Curtis breathed.

Kidder waved one hand in greeting but Stapleton did not respond. He chugged along without letup until he was close enough to speak. Then he offered one of his thin smiles.

"Pretty near dinnertime?"

Curtis pushed back the cuff of one mitten to discover the glittering oblong watch on his left wrist.

"Ten to twelve," he said.

240

"Your stomach keeps goddamn good time!" Kidder snorted.

Stapleton was busy unhitching the snowshoe clips.

"It don't mis-*lead* me too often."

All of us were waiting for some word of what he had discovered. He must have been all the way to the border and back. Yet he offered no report at all until Kidder finally asked him: "Any signs?"

Stapleton gave his head a scanty shake and seemed to size us up with his icy clear eyes.

"Just that old slote where somebody come down before the thaw. See where he went down Seven Ponds Stream. Headed for Oquossoc. Must have been Bill Jones. Long legs, anyway."

"That was Bill," said I. But Stapleton paid no attention. He seemed to be studying Kidder's face.

"Kidder," he said abruptly, "you know who's been using that old camp in the notch? Just this side of the line?"

Kidder shook his head.

"That was where Crocker stayed when he was working on the phone line. Nobody laid title to it since that I know of. Some of them border patrol jokers may take their women there." Kidder laughed extravagantly but Stapleton merely nodded.

"Well, someone's been keeping it up pretty good. Plenty of firewood inside, roof jack looks new. But nobody's been in it since the snow."

"Maybe some of them hairy old rumrunners," Kidder said.

"No, no. They got more sense than that. Once they get this side of the line they keep going."

"Poachers?"

Stapleton shrugged.

"Skedaddlers?"

Stapleton chuckled.

"*Might* be one or two of them left. Be kind of peaked by now, though."

Skedaddlers, as almost everyone thereabouts still recalled, was the name for Civil War draft dodgers who had holed up some seventy years before in a spot on the Kennebago Lake shore that was still known as "Skedaddlers Cove."

At this moment we were startled by a plaintive wail that sounded for a moment like the magnified bleat of a nanny goat. Curtis started as if he had heard a shot.

"What the *hell* was that?"

"Dinner," said Kidder. "Old Jenny found that horn they use to call in the help at chow time."

"Sounds awful good to me," said Stapleton. He fitted his snowshoes together to carry in one hand and waited for us to start up the icy path to the kitchen. "No need to say anything to Jenny about that camp up there. Just get her to fretting. But it wouldn't hurt to check that place when you get time."

Jenny stood waiting for Stapleton with her face drawn tight.

"You find anything?"

"Not a thing," said Stapleton heartily. "Nobody been within a mile of here except Bill Jones. I see where he went down the river."

Jenny nodded but seemed to take little cheer. She brought steaming platters to the table: ham, boiled potatoes in their skins, limp cabbage, turnips cut into quarters. She carried the big pot around to pour out coffee.

"Anybody want water?"

Stapleton, his mouth already full, grunted and pushed his tumbler forward. When it was full, Stape lifted it to his lips and quickly drained the glass. I heard young Curtis suck in his breath. He was staring at Stapleton in horrified fascination, as if Stape had just swallowed a glassful of spiders. And when Jenny made to fill *his* tumbler, he hastily turned the glass upside down.

It began to snow while we were eating dinner, lightly at first, just random flakes spinning down like bits of fluff from a shaken bedspread, then, the next time we looked, dropping dense as rain, swiftly and straight, for there was almost no wind. By the time we were ready to get back to the ice, the snow was already inches deep along the path, making for slippery going. There was only about two hours' work ahead of us — a dozen more cakes to cut and stow, and then a thick layer of sawdust to toss on top. The sawdust, mostly in frozen chunks, we flung in helter-skelter, to thaw and settle when the weather turned. When we were done finally, Kidder dug out brush from the shore and cut an armful to stick in the open hole to warn people away. By now we all wore a layer of soft snow on our hats and shoulders and we had to kick our way up the path to the kitchen.

Stapleton meanwhile had made a swing up the mountainside, looking for tracks and then headed down the road for Kennebago, after promising Jenny he would "keep poking around."

But Jenny, when we stomped into the kitchen, did not seem much comforted. It was nearly four o'clock and twilight was gathering. The snow, freezing now, whispered endlessly at the windows. I know Jenny dreaded the night, although it seemed to me that there was no chance at all for anyone to approach the camp in stormy darkness. We left our boots by the door, shed our coats and hung them on chairs, then gathered instinctively by the cookstove, where the blazing heat set our heavy wet trousers to steaming.

The unforgettable aroma of a camp kitchen enveloped me: the mingled odor of wood smoke and hot metal, the savor of fresh coffee, the pungence of wet wool, and the nostalgic, attic smell of old unpainted boards. Nobody spoke at all for a while, for the embracing warmth of the kitchen had lulled us promptly to silence. Wood snapped in the firebox and the tea kettle bounced in rhythm. Jenny hefted the big coffeepot and gave us a questioning look.

"God, yes!" said Kidder. "It's starting to come off cold out there."

"Bound to," said Jenny. Solemnly she set the cups in a row and filled each nearly to the lip. When the cups were full she turned to Kidder. "I tell you one thing, Mister Jim Kidder. Maybe it won't bother you any. But the boys is going to get awful tired of smoked ham before too many weeks goes by."

Kidder chuckled.

"It's awfully nourishing."

"It can get awful damn tiresome too," said Jenny. "Especially when you get it three times a day. I'm not complaining. But I bet these boys will be looking for something not quite so salty. After a spell it will draw the juice right out of your mouth."

Kidder seemed to ponder the problem for half a minute.

"How about if I work in a little venison?"

Jenny looked aghast.

"You mean jacking? Oh no! Please, Kidder! For God's sake! No shooting! Please! You'll have Eben Harnden and a slew of deputies in here next day!"

"Come on, Jenny. How the hell is Harnden going to hear a shot from thirty miles off? They ain't a warden between here and Rangeley. Even if there was we'd have the carcass hid before he could make it halfway up the trail. In this weather? Christ, he wouldn't stick his nose out the door."

Jenny sighed and shook her head in what seemed like utter misery.

244

"Well, I don't know where you think you'll find a deer this time of year. They been yarded up for weeks."

"They's a bunch of deer winters in that cedar swamp just up over the ridge, on the town line. Every winter. Me and the boys can start one out in a matter of minutes."

Jenny still shook her head unhappily.

"It's just that I don't like any shooting at night. Just never did care for it."

Kidder patted her shoulder.

"So you just sit tight and keep your ears plugged up and count to five hundred. We'll be back with liver for supper."

Now Jenny looked truly aghast. She grabbed Kidder's sleeve.

"Now wait! You ain't going to leave me here by myself? Oh, no! It's scary enough here in the daytime. I ain't going to sit here like a dummy waiting for some son of a bitch to come *shoot* me!"

Her voice cracked on the last word so it became almost a scream. The tears poured in a sudden torrent on her cheeks and she grabbed up her apron to stifle her sobs. For the first time since I had known him, Kidder seemed totally undone. His mouth fell open and he could not utter a word. He reached out and took Jenny by both arms, while she kept wiping at her cheeks to sop up the puddle of tears.

"Okay. Okay," he murmured finally. "I'll leave Bob on guard. I even got a gun he can use."

Jenny heaved a mighty sigh and managed to get some words out.

"I got a gun. Loaded too!"

"Jesus! You was ready for bear! I'm glad I got your girlfriend out of here before you took to ventilating her."

Jenny managed a sort of choking laugh and patted Kidder's arm.

"I'm all right," she whispered. "You go ahead."

"You sure? We can make it tomorrow night, just as well."

"I'm sure. If you're going to do it anyway, why you might as well get it over with."

Kidder kept his hand on her shoulder and bent to look into her face.

"Bob'll keep good watch. Not that anybody's going to be crazy enough to try to get out here in a blizzard. Won't take but a minute to start a deer out of that swamp with old Gordon thrashing around. We'll be back before you miss us."

Gordon, still nursing his coffee, sat suddenly erect.

"Well, listen," he said, "why don't I stay here and let Bob go? I am not too good with a rifle."

"Don't need to be," Kidder assured him. "You just come charging along on those barrel staves of yours and them deer will scatter. You brought them along?"

"Yes, but I'm not so sure about this. If it's against the law . . . Well, it might, well, you know, sort of queer me in college. And if it got connected with Dad . . ."

Kidder had closed one eye in an expression of extreme skepticism. But now he merely chuckled.

"Okay," he said. "You can stand guard over our fair lady and Bob and me will go out and fight the elements. Right, Bob?"

"I'm ready," said I, and grabbed up my boots. The spookiness and the sniff of danger already had my pulse leaping. When I crouched on the porch, however, trying to buckle the half-frozen straps on my snowshoes, with the snow sifting down like rice beyond the porch roof, the great empty darkness ahead caused my soul to shrink. The warmth of the kitchen soon drained out of my skin. I buttoned my mackinaw tight under my chin and pulled my cap well down over brow and ears.

Kidder had fetched a long flashlight and I was bringing the rifle — a handsome little carbine that weighed no more than an axe. I had to snuggle it close to keep it from slipping out of the choppers' mitts I wore over my wool mittens. The flashlight beam at first found nothing but the falling snow, straight as a beaded curtain, but soon after we had climbed out of the camp dooryard the yellow beam picked out tree trunks and thickets of tiny fir. The soft snow yielded quickly to our tread, so that before I had gone a few yards there were several handsful of snow riding in the webbing. There was no real wind and the chief sound was the tireless hissing of the falling snow. But occasionally the air seemed to shift with a sort of faint sigh, like a man moving in bed. I could see now that we were following a narrow trail or road that led us up the first gentle slope of the ridge. The birch that had marked our way at the start soon became black spruce and fir, the boughs so heavily festooned with snow that they seemed like a single canopy, scalloped irregularly as the tree line at times seemed to back away from the beam of light.

The trail began to climb more sharply and Kidder stopped after a few minutes, to blow, to shine the light on me and ask me, in a subdued voice, how I was coming.

"Good," said I.

"Tain't far," Kidder panted. "Kind of tough going, though. Be a hell of a lot worse without the snow. Brush is thick through here." The snow of course, some four feet deep by now, allowed us to walk free across the brush tops. The tiny trees we had seen along the way were actually the tops of young fir. Kidder waved one arm, sending the cone of yellow light swooping wildly across the curtain of forest. "Only about a half mile to the swamp. We hit the blue line once we get over the ridge here and that takes us right through the swamp." The smoke of his breath mounted through the yellow light. "Thicker than hell in there. Lots of moss for the deer to feed on and the snow don't gather.

We ought to spook one out of there right off. Then the poor bastard can't get any footing in the snow. They really ain't made for this country."

He started off then, chugging steadily, and I labored along behind, just inside the rearward glow of the light. As we reached the top of the ridge, with the falling snow blinding us to what lay ahead, the tree closest to me exploded. A bomb of snow struck my head. Sudden thunder nearly deafened me. I jumped, in spite of myself, got my snowshoes crossed and very nearly plunged into the drifts, rifle and all. Kidder stopped and held the light on me until I had recovered my balance.

"Goddamn partridge!" he yelled. "Son of a bitch ought to of been under the snow where he belongs!"

I did not feel like admitting I had not recognized the sound of thundering wings. I cursed dutifully, straightened my snowshoes, wiped my face with a mitten, and shuffled on.

The swamp, when the light fell on it, looked like the open entrance to a tomb. There was a black tunnel where no snow fell, a floor from which a few snow-crowned stumps protruded, and dark trunks that crisscrossed like sagging rafters in a collapsing barn. And it was quiet, quiet, quiet as a haunted house. There was the suggestion of unending darkness beyond. Kidder stopped, as if he needed to take a deep breath, or even say a prayer before he committed himself to the unknown. Then, before either of us could move, a scream split the silence, like the sudden whine of a buzz saw. A buck had challenged us. Kidder began to inch the shaft of light across the swamp, showing one dark opening after another. Then two small answering lights glowed back, golden bright, motionless, like distant headlights. Kidder reached back, motioning for the rifle. I shuffled close to pass it into his hand while he held the deer transfixed by the flashlight.

"I'll hold the light," I whispered.

"No, no." Kidder spoke in a normal low tone. "Shooter has to hold it."

It seemed to take him minutes to get light and gun adjusted without moving the spotlight off the deer. Finally, looking over his shoulder, I could see the tiny gleam of the front sight. Instantly the rifle shot cracked the night open. The spout of orange flame seemed for a fraction of a second to blot out the world. Then the sound was smothered in the snow and arching trees, with only the faintest echoes trickling back from nowhere. Kidder was already plunging into the swamp with the light leading him. I hurried after him, to stay within the glow.

The deer lay on the snow, front legs bent beneath him and antlered head bowed between his knees, as if he were begging forgiveness. I wish to Christ we hadn't done this, I told myself.

Kidder shoved the flashlight into my hand and moved my hand to center the light on the deer's flank.

"Hold it there! Right there!"

He set his rifle daintily across a stump, took his knife out and cut a small branch off from somewhere in the dark. Working in the light of the flash, he shaped the stick with his knife until he had a long rod with a hook in the end. He handed that to me. "Hang on to that," he said. He squatted then to unbuckle his snowshoes. The snow here was no more than a foot deep. Kidder pulled the deer around until the buck lay flat on its back, forelegs to one side and the hind legs held apart. Working deftly as a surgeon, Kidder cut the long penis free from the abdomen, the whole length, then flung it aside so it hung off the deer's flank in the dark. It seemed almost half a yard long. Kidder cut off the testicles next and tossed them off into the snow. With the knife in his right hand Jim put his left hand on the deer's belly, put the tip of the knife between the first two fingers of his left hand, then, pressing the abdomen down with his left hand, he started to cut, following the blade along with his left hand to make sure the knife did not cut into the membrane that enclosed the paunch. The guts rolled out in a glistening sack, threaded with tiny dark veins. Kidder slashed at the diaphragm and it opened with a quick sigh, like an indrawn breath, marking, I thought, the soul's escape. Then Kidder rolled up his sleeves and plunged both arms, knife in one hand, far through the deer's chest to find the gullet and cut it loose. For the first time blood began to run in a stream that spread, black as muddy water, to melt a gutter in the snow. The bullet wound in the head had not run blood at all. It was a neat round hole below and between the antlers that must have killed the animal instantly.

Kidder found the bladder and pressed out the urine so it ran in a silent stream from the penis.

"Get that piss on the meat and it ruins it," he said, as if I had questioned the move. Next he sliced all around the breech to free the lower end of the bowel. Now he could tumble the whole gut out into the snow. He fingered the liver free, sliced off the gall, and handed the liver to me, all a-glistening.

"Hang that on the stick. That'll be for supper."

He used the knife to cut all the membranes that held the chest organs in place, and rolled heart and lungs out at my feet. The guts steamed gently, emitting a heavy odor that was almost sweet.

Kidder washed his hands painstakingly in the snow, hunched into his jacket, and absently nudged the steaming guts into a heap with one foot.

"Leave that for the foxes," he said.

He fished a length of rope from a jacket pocket, made a

loop in it, and dropped the loop over the deer's antlers. He squatted down and buckled on his snowshoes.

"You go first. Just don't drop that goddamn liver."

"I'll give you a hand."

"No. No. That won't work. You just light the way. This critter don't weigh much over a hundred. And once we get over the rise here, it's all downhill. Go slick as a sled. Just stay in the slote and you can't go wrong."

The trail we had made was easy to follow even though new snow had blurred it some. The going seemed easier too and before I realized how close we were my light caught the corner of one of the camps, stained logs and a dead window that threw the light back.

"Whoa up!" Kidder called. "We'll hang this here in the woods."

I brought the light back and held it while he flung the rope over a maple branch and hoisted the carcass until its hind feet were clear of the snow. He stuck a stick in the incision to hold the body cavity open, then dug out a bandanna from somewhere in his clothes and tied it around a forefoot of the deer.

"That'll keep the cats away," he explained. "Or at least it'll give them something to think about."

In the kitchen we found Curtis and Jenny sitting almost knee to knee before the stove, where kettles were steaming. They looked back at us, half smiling, as if we had interrupted a story.

"Liver for supper!" Kidder announced. I held my trophy high.

Jenny cried out like a child at Christmas.

"You got one! Already!" She jumped up to find a pan for the liver. "We never heard the shot."

Curtis, still in his seat, was studying the gleaming liver with obvious distaste. "Are you going to *eat* that?"

"*I'm* not," said I.

Curtis spoke up quickly.

"Well, I certainly won't either. I swear I simply can't *go* that stuff!"

Jenny, having taken the liver in a tin plate, set it abruptly on the counter by the sink and dropped both arms to her sides.

"Well, Jesus to Jesus!" she cried. "I sure ain't cooking this stuff if nobody's eating it!"

"Don't you fret one minute," said Kidder. "I never said no to deer liver."

"Me neither," said Jenny. "So the boys will have to settle for ham." She reached out and gave Curtis a light shove on the shoulder. "I don't know what we're going to do with *this* one. He won't eat the meat and he won't drink the water."

Kidder gave Curtis a puzzled glance.

"The water? What the hell's the matter with the water?"

Curtis shrugged and shook his head.

"Oh, I don't know. It's just . . . all the crap in it."

"What crap? You want to know about crap? What do you suppose all these club members was drinking for years? Beaver shit! Why me and Judkins was up here to the spring one fall. Up where they set that intake. It's more of a brook there, where it makes that turn. The beaver put a dam in there years ago." He looked at Jenny. "You ever seen that place?" Jenny shook her head. "Well, right by that dam they's a real deep hole and that's where the club set that intake. Well, Jesus Christ! That old grandpa beaver, he'd been using that for a toilet!" Kidder shook his head and laughed. "They was little piles of these old droppings all around the hole. Christ only knows how long the club had been drinking that water! Imagine what a shit fit they'd have had? If they'd knowed? Well, me and Judkins, we knocked that dam out and let it flush out good. Then we raked out all them droppings and cleaned the place all up. Course we never said nothing to the members."

Curtis was staring at Kidder in open horror. His face had gone pale.

"My God!" he gasped. "Does Dad know about this?"

Kidder squinted at Curtis for two or three seconds, probably realizing for the first time that he should not have told his tale in this company.

"No, he don't know about it. And he ain't *going* to know about it!" All the laughter had gone out of Kidder's tone and his face was grim. "You say a single goddamn word to him, or to *anybody* in the club and I . . ." Kidder worked his lips and swallowed, as if his anger had been something solid. "Well, by Jesus, you'll be off my list." The limpness of this threat obviously annoyed him. He leaned forward and glared into Curtis's face. "Don't say one goddamn word to your old man! Not *one goddamn word!*"

Curtis's mouth trembled as if he were about to cry. His face had gone suddenly red, then pale again.

"Well, all right. But my God! All those years!"

"Them years is all gone. The spring is clean as it needs to be. And nobody got sick from it, that I know of." Kidder rapped his hand on the table. "Jesus! I never should have told you that."

Curtis shook his head solemnly.

"Well, I'm not going to *say* anything. I swear!"

"Well, that's good," said Kidder. But he was obviously not entirely content. For the next two days, as we labored together out of Curtis's hearing, shoveling the crusted snow off the cabin roofs, he would stop at times and vow bitterly that he should kick himself in the ass for ever having told that goddamn story. The sun was bright those days, glittering relentlessly off the snow and setting the icicles everywhere to dripping. Then, at night, with the sky still shorn of clouds, the moon would stare innocently down from a universe of stars, as if there had not been any storm at all. Night or day there was never a sound. No animals moved. The wind lay still. Darkness came at half past five and saw all of us sitting in the kitchen, offering our unshod feet to the cookstove or bent over the table cuddling our teacups in both hands.

Curtis had been set to the job of opening pathways to the woodshed, the hovel, and the camp where we slept, keeping the horse's crib full of hay, and fetching water from the pond. This chore enabled him to make frequent stopovers in the kitchen, where tea and molasses cookies were always on hand and where, I did not doubt, he and Jenny were sharing intimacies I could only guess at. Once or twice, when we were all in the kitchen, I had an impulse to put an arm around Jenny's waist, to establish my priority. But at that stage in my life this was a gesture far too bold for me to initiate. So I registered my general disapproval by occasional spells of glumness, which I am sure no one ever took note of.

In the cabin one night, Curtis, aglow with a good supper and with goodwill, allowed to us both that Jenny had "a lot on the ball." Although this phrase, in that day, was usually meant to indicate, in a man, transcendent skill at one's job, an abundance of social graces, or a gift for repartee, applied to a girl it usually suggested sexual attraction surpassing mere physical charm, or, even, when offered with the appropriate grimace, special talents at lovemaking. I bristled, therefore, at the remark and probably emitted a grunt. But Kidder laughed.

"You just find that out? That's some special lady. But she's too goddamn old for you, laddie."

Curtis offered a gracious chuckle.

"Oh, I don't mean *that* way. But you know I bet she was hot stuff in her day. She's *still* got lots of what it takes. I

254

mean for a woman her age. What is she, thirty or something?"

"Right around there. Maybe a few years short. Real gone
by. Like me."

"Oh, come on now. I mean it's different with a woman.
Didn't she ever get married?"

"Hell, yes! She's still married. You go smelling around
there and old Myron's liable to fill your ass full of boots."

"But she said . . . I mean, she lives by herself, doesn't
she?"

Kidder winked at me.

"Oh, yes. 'Cepting when Bob's holding the fort."

Curtis turned to me with a puzzled expression, then
looked back at Kidder.

"Are you horsing me?" He turned to me again. "Is he
serious?"

"Don't believe everything he tells you," I said. But I rejoiced, all the same, at the turn this talk had taken and I wondered what words I might find to hint at the truth without
making myself out a braggart. But it was instantly clear that
Curtis was going to believe Kidder anyway. He was still
looking at me in mild astonishment. Then he turned back to
Kidder.

"You mean, he *stayed* with her? Stayed in her house?
Wasn't her husband even there?"

Kidder could not stop laughing.

"Hell, no! Jenny wouldn't let Myron within a mile of the
place, less'n he come on his hands and knees. She ran him
the hell out a year ago. Then she found Bob wandering
loose without no place to lay his head and she took him in.
What the hell else she did with him I couldn't tell you." His
laugh now filled the cabin. "We'll have to give Bob some of
that truth syrup, or whatever the hell it is, to find out what
went on in that place."

Curtis made an effort to laugh but he seemed truly aghast.

"You really *lived* there with her? Jeepers!" He studied my face as if he might find there some mark of villainy or shame. "Weren't you taking a hell of a chance? I mean isn't there a law? I mean not being married. I mean living together and all that?"

"Jesus, Gordon!" Kidder laughed. "You must know by now they ain't no law above Bemis."

Curtis again made an effort to laugh but the noise was far from happy. He continued to study my face intently.

"Well, I know. But my God!" He squinted to focus his eyes on mine. "Did you *screw* her, for God's sake?"

"No!" I exclaimed. The word came out without any thought and I never did know exactly what the secret promptings were that impelled me to deny the truth. But I did not regret it then or now.

Curtis just sat with his mouth open for a moment, not sure if he had been the victim of one of Kidder's wild fairy tales.

"But you did *live* there?"

"I just had a room there. Just for a few days while I was waiting to come here."

Curtis relaxed now and managed to offer a slightly sheepish grin to Kidder.

"Jeepers! You had me going there for a while. I thought . . . Well, you know . . . I didn't know what to think."

Kidder's laugh had subsided to a steady chuckle.

"Think about this one, then. They ain't no spare room in that house."

"Oh, come on!" said Curtis. "I'm going to bed before you get me *all* bollixed up."

❦ 9 ❦

Four days after we had shot the deer Jenny told us at breakfast:

"I wish you'd skin that critter and get him out of sight. Stapleton will be back here any day and I just as soon we don't have that hanging in his face."

"Oh, Stape won't give a damn," said Kidder. "He's got bigger fish to fry. But if old Harnden takes notice of all this traffic in and out he might decide to ride in. Be just like him. Time we started feeding on that carcass anyway."

Curtis walked out with us to see where the deer hung, the hide all frosted with snow. But he held back as Kidder made ready to cut into the hide. He wanted to be able, I suppose, to distance himself from the crime, if the law should suddenly descend. And when Kidder, having run his razor-sharp knife around the animal's neck, had begun to work the pelt free, Curtis suddenly observed that there was a good ski run nearby, down open snow, clear to the lake, so he went shuffling away and, pushing himself off with the pole he carried, seemed to take flight over the snow, swooping down and up like a gull, powdered snow rising like road dust behind him. Kidder and I both stood staring. I had never seen skis used this way before.

"Jesus!" said I. "He's good!"

Kidder laughed.

"He's got to be good at something. He's been at that since he was in grade school. Too bad he can't make a living that way."

"He won't have to."

"That's a fact. Unless Daddy takes a dislike to him."

We watched Curtis's flight until he had come to a stop finally, far out on the lake, yards beyond the brush that marked the ice hole. He turned and seemed to skate back, shoving only occasionally with his poles. Before he started to climb the slope, he looked up at us, his face bright as a happy child's.

"Nice going!" Kidder called.

We went back to our skinning, with me helping occasionally to pull the heavy hide free, using one forearm to push down on the loose skin while yanking it free with the other hand. Kidder kept slicing the thin blades of membrane that held the hide to the flesh, so that skin came free unblemished. There was a primitive satisfaction in taking the hide off whole this way — like peeling bark unbroken from a tree. When the job was done — the hide cut loose at the hooves and hung to dry — Kidder chopped the head and feet free of the carcass and we carried the body, naked as a new baby, into the shed to butcher it.

Kidder split the carcass across the middle, then cut out the saddle, leaving forequarters and hindquarters whole. These he hung from the single rafter in the shed and brought the saddle to the kitchen, where he found a cleaver to help him separate the chops.

"This'll eat as good as anything you ever tasted," he told me. "This was a young buck, tender as a lamb."

He took over the cooking at dinnertime, frying the double chops in a smear of bacon fat and salting them gently as they sizzled in the spider.

"A lot of them old bastards won't eat deer meat until it's been cooked to a frazzle. Gray right through. They never did learn how good it could taste."

Kidder served the chops cooked pink, like fine lamb chops, and even Curtis, who had at first picked at the meat as if it might turn out to be full of worms, began to feast on it heartily, with constant small expressions of delight. And I had to agree out loud that it was indeed as good as anything I had ever tasted. Better certainly than anything that had appeared on my plate since early the previous spring.

After dinner, we buckled on our snowshoes, picked up our shovels, and set out to shovel more snow.

"We'll go up and do that Beasley camp," Kidder told Jenny. "That roof ain't going to hold much weight."

"Away up there?" Jenny seemed truly puzzled. "They still *use* that place?"

"Well, somebody might. And my contract is to clean off all them private camps. Nobody said I should skip that one."

"I didn't know Beasley was still alive."

"Maybe he ain't. But if we don't get that snow off pretty quick he won't even have a house to haunt."

"You got a long trek."

"That don't bother us. Right, Bob? Us old woodsmen can go from dawn till dark, eight days a week."

"As long as we have deer meat to stoke the fires," I said.

But we were hardly out of earshot when Kidder, after a glance back at the kitchen, told me:

"We're going up to check that linesman's camp that Stapleton looked at. I want to find out who the hell's been using that. And I don't want old Jen to get her liver full of hot piss when there's no need."

It was three miles to the border, up the length of the lake and along an old two-sled road that, with the brush buried, seemed wide as a highway. The cabin we were looking for stood well off the road, in a small clearing. But before we ever reached it, we found the snowshoe tracks. They were not entirely fresh but clearly indented and crusted over where the sun had melted the snow surface only to have it freeze over in the night.

"Stapleton?" I suggested.

"Hell no. That's since the snow. Somebody come up from downriver and made straight for the cabin. Maybe a poacher. People *will* go shooting deer out of season, no matter what the good book says."

We followed the tracks along and found fresh tracks leading away from the cabin to the north.

"Heading for the gate," said Kidder.

The roof of the cabin was gleaming black, all clear of snow except the very eaves, where icy snow still gathered and icicles began. Kidder studied the scene from a rod away.

"Must have stayed here awhile from the looks of the roof. Moved around some too."

There was frozen snow on the cabin step with the shape of footprints showing and two slots where snowshoes had been planted for the night. Footprints marked the trail to the tiny outhouse, shoveled free of snow to allow the door to open.

"Fussy son of a bitch," said Kidder. "Wouldn't shit in the snow." He moved closer to the cabin, mounting the snowdrift and bending low to peer into the window. The snow reached to the sill.

"Lot of wood in there," he murmured. "A few cans of

grub. Looks fresh too. That'll freeze solid, less'n he plans to get back soon." Kidder shaded his eyes with one mittened hand and peered more intently. "For Christ's sake," he breathed. "That's no poacher. Left his Jesusly rifle behind." He backed away from the window, took a look around, and pursed his lips as if he might whistle. "Jesus. You know who that is? That's the Jesusly warden. Eb Harnden. He must be the one's kept this up. Let's fly to Jesus out of here. We got work to do!"

Kidder painstakingly lifted his snowshoes to get them reversed, then he started swiftly back along our tracks. He hit full stride immediately, the big shoes flinging gobbets of snow right and left. By the time I got myself turned around, after getting into a slight tangle when I set one snowshoe atop the other, Kidder was yards ahead. I saw that he still carried his shovel and I had to reverse myself and go back to get mine from where I had planted it in the camp flooryard. Kidder waited for me at the lakeshore.

"We got to get that Jesusly carcass out of sight!" he panted. "He's going to be right on our trail, soon as he gets back."

Kidder set off again and soon led me by a hundred yards. He slowed down after we had gone about a mile across the barren lake but he did not stop, nor could I ever catch up. He was out of sight when I reached our own shore but I knew he had gone straight for the shed. Jenny and Curtis were standing side by side on the kitchen porch.

"My goodness," Jenny cried. "You fellers seen a ghost? What's all this tearing back and forth."

"Tell you in a minute," I panted. I moved right along on Kidder's tracks and found him already manhandling the frozen hindquarters of the deer. "Grab that other junk!" he shouted. "Bring it along."

With no notion at all of where we were headed, I took up the forequarter piece, which seemed surprisingly light, and held it to my chest as I chugged along after Kidder. Curtis had come out on the shoveled path to gape at us.

"What's going on?" he cried. Kidder, already far out on the slope, turned to face him.

"You come too!" he yelled. "Get them Christly barrel staves on!"

Curtis turned to me.

"What's this all about anyway?"

"Don't ask me," I replied. "I'm just doing what Kidder said." I plugged on then to catch up with Kidder. He stood at the top of the slope, looking down over the stretch that Curtis had sculpted with his skis.

"Now look," he said. "You just stomp along behind me. Set them shoes in hard. And after I throw this junk down, you hand me the other. Stay right behind."

I did as I was told, setting my snowshoes down firmly with each step. After Kidder had gone about ten yards he raised his share of the deer carcass high above his head and slammed it with all his strength into the deep snow at his feet. It sank instantly into the fluffy drift, leaving only a tumbled indentation, which Kidder immediately tamped down with his snowshoes. He turned then and took my section, plodded ahead a few more feet and slammed it into the snow.

"You come on now," he charged me. "Walk right on over." So we both trod the snow down over the burial site and continued on to the lake, where our tracks mingled with the half dozen tracks we had made in our comings and goings. Curtis had appeared by now, sliding easily along on his skis.

"What's this all about?" he called. "What'd you do with the venison?"

"We put it in the snow. It's going to keep better," said Kidder. "And them animals ain't so liable to find it. You go on up and slide down over our tracks and firm it up good, so them foxes won't start digging." Kidder turned to me then, apparently not sure I would accept this nonsense as the solemn truth, and dealt me a grimace and a wink. But Curtis moved eagerly up the slope and soon came swooping down, holding the poles tucked under his arms, graceful as a dancer.

"Hey!" he cried out, looking happily into our faces. "I swear! That's the best damn slope I ever tried! That's really the berries!"

He set out at once to climb back up the slope, toeing his skis out and clumping ahead with the gait of a circus clown. Kidder and I watched in silence as he positioned himself high on the slope and came streaking down again, more swiftly this time, so that the momentum sent him zipping far out across the ice. He turned and skimmed back toward us.

"You fellows want to try this?"

Kidder laughed.

"Jesus, no! I'm too old for that crap!"

"Come on! It's really a snap, once you get the hang of it. You want to try, Bob?"

I did want to try but I was too uneasy about the warden's arrival to leave Kidder's side. So I shook my head and we watched Curtis for another ten minutes, keeping watch too on the lake for the dark form of a man coming our way. But no man came. I kept covering my mouth with one mitten to blow warm breath across my nose and cheeks.

"He ain't got back yet," said Kidder finally. "I don't know what the hell he'd be going out to the gate for. Maybe he just swung around through the woods to take us from behind." He laughed at this notion. "But shit! That'd just take him two miles out of his way. By that time our boy will have that deer meat packed down so's he'd have to dynamite. Only . . ." Kidder stiffened suddenly, as if he had been poked from behind. "Jesus Christ! We got to get the fuckin antlers out of sight!" He turned and labored up the slope, blowing noisily. I hustled along behind.

The head and antlers had been dropped in a corner of the shed, where they had frozen to the floor. Kidder grabbed them up, leaving a small patch of hair and frozen blood still clinging to the boards. He scuffed at this with his heavy boot and scattered some crumbs of bark on the spot. "Dassent take these out on the slope," he said. "Old Gordon would have a shit fit."

"He'd want to hang them on the wall," said I. Kidder looked at me with wide eyes.

"By Jesus, *that* ain't such a bad idea. The antlers is all that counts anyway. You fetch me a clean board from over in that mess there, and I'll do the rest. We could hang these up and old Eben wouldn't know but what they'd been here forty years."

"Unless it starts to thaw."

"Won't matter. Won't be anything but bone. And I could stick them up in the dining room, anyhow. Colder than a witch's tit in there."

He had taken up the blind head and positioned it on the bench in front of him. With a small hacksaw he set out to cut off a section of the scalp that held the antlers. It came off neat as a piece of store-bought crockery and Kidder nailed it easily to the board I gave him. It fitted squarely in the center.

"Keep an eye out for old single-tit Eben while I sneak this into the dining room," he muttered. "Speak up if old Jenny starts to get nosy." He wrapped his burden clumsily in a bit of torn and frost-stiffened tarpaulin and dodged away on the shoveled path. I walked out to the porch of the lodge and stood pretending to study Curtis as he careened down the long slope and labored breathlessly back. He seemed each time to fly farther out over the ice, sometimes gaining a few extra yards by shoving with his ski poles. The sun was high now, shortening all the shadows, and still no Harnden appeared, nor any moving figure at all, on the long lake surface, where a million specks of sequin glittered from the snow. Kidder came up in a moment and invited me to look through a window to see where he had hung the antlers over the dining room door. In the half-light there they looked like just one more of the dozen or so dusty trophies that adorned all four walls.

"All we got to do now," said Kidder, "is get rid of that Christly deer head. You go up to the camp and get a good fire going in the heater stove. This'll fit right in now and it'll burn to ashes. May stink some. But I'll tell him we're just burning your dirty underwear, case he notices."

"What about the hide?"

"I got a place for that. I'll just take it and hang it in the Crosby camp, behind that Indian blanket or whatever he calls it that he's got hanging on the wall."

But when Kidder came back, carrying the bloody head in the remainder of the tarpaulin, he did not seem hurried at all. I opened the stove top and he looked at the new flames, but made no effort to consign the skull to the fire.

"Need another stick?" said I.

"Uh-huh. But you know what I think? I think we been wasting our time. I don't think Harnden was up there at all. I've been thinking about the cabin. There was something funny about that place. I noticed it and I didn't notice it. You know how you do sometimes? You see something but you don't really pay it no mind until later. Well, I've been thinking it was damn funny, the guy staying there all night but there wasn't no pee hole in the snow. Now I don't give a damn how young a man is, when he gets up in the morning he's got to piss. Right? Just about the first thing you do when you get out of bed. And old Eben probably has to get up twice in the night. Well, this guy didn't."

"Maybe he pissed in the cabin."

"Oh balls! Maybe he pissed his pants too. No, that has to of been a woman. No two ways about it."

"A woman, for Christ's sake? Who?"

"Well, she come up from Little Kennebago way. Had to be pretty rugged to make that on snowshoes, so it wasn't one of them little flipper flappers that slurps up them ice cream sodas at Fowler's. Well, I don't know. You got any ideas?"

I didn't know any women in Oquossoc outside of Jenny that I would have called rugged. I knew hardly any women at all. The postmistress. Her plump sister, Mrs. Fowler, who wore full makeup and high heels every time I saw her. Scrawny Mrs. Hill, who was too sick to walk across the road. Kitty Dunham.

"Kitty Dunham." I spoke under my breath, lest Jenny be too close. And I set a new stick of maple on the fire.

Kidder nodded. He dropped the deer head into the stove, where it sent sparks flying. I closed the stove top.

"That's just what I was thinking. But Jesus Christ, don't breathe that to Jenny. As far as she's concerned we're just worried about a warden. And maybe Harnden really was up there. Maybe he pissed the bed and went out to the gate to get fresh underwear. But I think we better sneak back up there and make damn sure. If that crazy old bag is up there with a rifle . . . Well, Christ knows what she mayn't do. She's crazy enough for anything when she's hot."

"Did you see any bottles?"

"No. But I wasn't really looking for them. Couldn't see but a small part of the cabin anyway. That would be *one* reason for her to head out to the gate. That place is just crawling with bootleggers. I think them immigration fellers wouldn't be above making a few dollars that way."

Jenny appeared then to tell us it was dinnertime. Then she stood with her face wrinkled up, trying to make sense out of our being there at all.

"What you fellers been up to anyway? You been racing around here like the woods was on fire. You was supposed to be shoveling off the Beasley cabin."

"We been hiding that goddamn venison if you want to know," said Kidder. "We picked up old Eben's track up above the northwest landing."

"Eben Harnden? He coming here?"

"Seems likely. He's smelling around for something."

"But Jesus. I got six of them chops on the table!" Jenny looked truly frightened. Kidder patted her shoulder.

"Don't fret. If we see him coming, we'll just dump them in the cookstove. But he ain't in sight now. We can get them all et before he's halfway across the pond. Bob, get our boy in here before he gets that slope wore down to the grass. We'll finish that Beasley camp this afternoon, if we ain't all been arrested." Kidder offered me a quick wink at this and followed Jenny into the kitchen.

But we did not visit the Beasley camp that afternoon or ever. We had hardly finished dinner when we heard the faint jangle of harness bells out beyond the knoll. We all looked at one another with what some poet might have described as a wild surmise. Then all four of us turned out in a rush to see what new disaster might be in the offing. The horse came promptly into sight, trailed by the storehouse pung, the very vehicle that had borne Tim McCormick's body on its final journey. Two men sat in the driver's seat — Jerry, the driver, and a man so deeply muffled in a fur-lined coat that his face could not be seen. He wore a bright red stocking cap.

"Jesus," Kidder whispered to me. "I know that goddamn coat. That's old man Curtis himself. He can't be up to no good."

Gordon Curtis, who had followed close behind us, confirmed Kidder's guess immediately by plunging, without thought, off the porch into waist-deep snow. The snowshoe tracks, which provided an illusion of firmness, gave way instantly under his feet, and he had to flounder back to the porch, plastered to the shoulders with snow, waving one hand the while. "Hey! Dad!" he cried.

His father returned the greeting with notable restraint, apparently reluctant to bare too much of his flesh to the cold. He lifted one hand about shoulder-high and then concerned himself with finding a safe place to alight. Jerry brought the pung close to the camp, where Gordon was able to help his father make the short leap to the shoveled path. The elder man was wearing regular lumberjack shoepacs, called "rubbers" in the woods in that day, and they showed never a blemish. His fancy cloth coat, lined with a fur that looked like beaver, fell almost to his toes when he stood up. He took hold of both Gordon's elbows.

"Well!" he cried, in a resonant voice that surely must have carried partway up the lake. "How are you, son? How do you like this life?"

"Oh, great! They have the greatest ski slope here, Dad. Really! It's absolutely tops! You know, I'd just as soon finish my vacation right here. If it's all right with you?"

The elder Curtis hardly seemed to hear. He patted the boy's shoulder as if he were dealing with a child who had run up to show him something he had made in school.

"Really? Fine! Fine!" he declared. "Where's Kidder?"

Jim Kidder was already almost at Gordon's side.

"Right here," he said, smiling broadly. "Glad to see you." The two men clasped mittened hands briefly. The elder

Curtis was a head shorter than Kidder. He had a round red face and eyes that seemed continually seeking some target they might impale. He fixed me for a moment with a glance that was intent and something short of friendly. "Who's this young man?" he demanded.

"Oh, he's my sidekick, Bob Smith," said Kidder happily. "Bob, this here is Mr. Curtis. The *big* boss."

"M-mmm," said Mr. Curtis, who touched mittens with me almost absently, then stepped up on the porch and looked around for some place to light. "You suppose I could find a cup of coffee around here?" At this instant Jenny, who had drawn back a few paces toward the kitchen door, stepped forward, her gentle face aglow with what seemed like undiluted happiness.

"You come right into the kitchen! They's a pot on the stove!"

Mr. Curtis acknowledged her with a sort of automatic smile.

"Do I know this young lady?" Apparently he addressed this question to the whole company. Gordon, who had hurried to stay at his father's side, took a step toward Jenny.

"Oh, this is Mrs. White, Dad. She sort of cleans up and cooks. And everything."

Jenny laughed. "Everything is right. They don't seem no end to chores around here. But Gordon's a great help!"

"Well, how do you do?" said Mr. Curtis. "Yes, I'm sure he is." He made his last remark sound as if he wasn't sure at all. "I'll just try some of your coffee then." He pushed forward quickly, his long coat nearly hiding his feet, so that he seemed to be moving on wheels. Jenny had to scurry to get the kitchen door open for him.

In the kitchen Mr. Curtis shed his amazing coat, looked about as if he half expected someone to take it from him, then draped it tenderly across an empty chair. Jerry, having blanketed his horse and found hay enough in the hovel to warm the creature's stomach, came in to join us for coffee.

"Fixin' to storm," he announced.

Mr. Curtis looked genuinely alarmed. He turned an almost angry gaze to Kidder, as if big Jim had had a hand in this new mischance.

"I'll just have my coffee then," he declared, "then Jim and I can complete our business. If you can take a few minutes, Jim?"

"Sure thing," said Jim. He shared a slight frown with me that seemed to say, "Now what?"

"I might have saved the trip," Curtis said, "but I couldn't get an answer on that damned phone. Doesn't that ring in here?"

"Line's down," said Jim. "Can't fix it till we get a thaw so's we can see the break."

"Well, it doesn't matter. Probably better this way."

Mr. Curtis, having received his mug of coffee from Jenny, sipped it gingerly and looked deep into the cup, apparently to assure himself it *was* coffee. We all sat around the upper end of the long table like participants in a ritual, silently communing with our coffee and waiting for Mr. Curtis to say his next say. He finished his coffee with a long draft finally, found a paper napkin to dab his mouth, nodded a sort of thanks to Jenny, and fixed Kidder again with his sharp glance.

"Someplace where we can talk?"

Kidder seemed really taken aback at this.

"Up here to my cabin," he said. "I'll just touch up the fire. You want Bob along?"

Curtis turned his gimlet gaze on me. But he spoke to Kidder.

"He with you on this job?"

"Second in command." Kidder grinned at me.

"Bring him along." Curtis turned to Jerry. "How much time we got?"

Jerry, who had been coping with a doughnut, had to waste a second in swallowing. He gulped and talked through half a mouthful.

"Can't tell. Looks pretty black in the west. Them clouds is full of wind."

Curtis drew in a tiny breath to indicate his impatience with this reply.

"Well, this won't take long. Where is this place?"

Kidder led the way out the door. Gordon stood up as if he might come along, but he did not move away from his seat. His father gathered up the heavy coat and draped it over his shoulders. He followed right on Kidder's heels.

The cabin was cold but there was still some glow left in the stove. Kidder aroused the flame with strips of birch bark and kindling.

"Now what the hell's all this rumpus about?" Curtis demanded. "That goddamn tinhorn sheriff sniffing around, yapping about some shooting or other. You got the club mixed up in this? Or the company?"

Kidder's neck grew instantly red and he turned a grim face to Curtis.

"I haven't got the club mixed up in a goddamn thing! Nor the company neither! What I been doing is my own personal business!"

"You know something about this shooting, then?"

"Well, Bob does. I know what him and Jenny told me."

Curtis, scowling and intent, turned instantly to me.

"What do *you* know about it?" His tone clearly implied that the chances of my knowing anything worth a moment's thought about anything were slim indeed. I found my throat almost too constricted to release more than a husky whisper.

271

"I was there when it happened," I said.

"When *what* happened?"

"Somebody took a shot at Jenny. It just missed me."

"Shot at her and missed you!" Curtis made it sound like an obvious fairy tale.

"Well, maybe shot at both of us. It was dark."

"And how did you know it wasn't a stray shot?"

"It was dark."

"So you said. And people don't shoot stray shots in the dark?"

"Well, maybe if they're lost." This remark sounded just as foolish to me as it must have to him.

"Exactly. And on the basis of this you started all this hoopla about someone trying to kill someone. You made an accusation against one of the best men in the area."

"I didn't accuse anybody." My ears and cheeks were burning painfully now.

"Well, who did then? That Toonerville sheriff? That man hasn't got the brains God gave a goose. Is he the one that started these rumors?"

I could only shrug feebly and shake my head.

"Well, I want it stopped!" Curtis was talking to Kidder now. "It's important to the club and to the *company* that we have a good relationship with our neighbors. With the *decent* people at any rate."

Curtis was shouting so now that everyone in the kitchen could have heard him. Kidder stood silent, his face gone white. Curtis wet his lips and went on in a much more subdued tone. "Who the devil is this Jenny?"

"Girl in the kitchen," said Kidder.

"I know *that*, for Heaven's sake! Is she married? Where's her husband?"

"She's married. He's in Rangeley."

"And she's up here by herself with three young men?

What the deuce kind of a deal is that?" He looked our cabin up and down as if he might discover some glaring evidence of unchurched sex. Kidder and I held our peace while the older man seemed to simmer down. He spoke now in his normal tone, which was still deep and resonant. "Well, I want her out of here. Right now."

Kidder shook his head. His lips were bloodless.

"I can't do that," he said.

To my amazement, Mr. Curtis's anger seemed to deflate ever so slightly. He drew in a deep breath, then proceeded in an even quieter manner.

"Why the devil not?"

"Because I told her she could stay here until . . . she felt safe."

"And you don't think it hurts the club to have this . . ." Curtis seemed unable to pick out a polite word for it. ". . . this *stuff* going on?"

"Nothing's going on," said Kidder. He seemed to be struggling to hold back all the words in his throat. His shoulders were actually trembling. Curtis studied him for a while and finally seemed to decide that the flaunting of authority was not going to do any more than delay the explosion. He made as if to brush dust from one sleeve.

He nodded five or six times and made a thoughtful face. All reasonableness and friendship now, he said very quietly indeed:

"Well, I hope you understand that you're not going to be doing yourself any good with the company if you persist in getting involved . . . Well, I mean you'd be very well advised to steer clear of this Jenny woman and this whole hoopla about shooting and what not. This fellow Kendall stands pretty high with us and the people in Berlin are not going to sit still and see him, well, slandered."

"I'm not involved in a goddamn thing," said Kidder, whose face still had not recovered its normal ruddiness.

"I'm glad to hear it. Just keep your nose clean. I mean stick to your routine and just do the fine job I know you're capable of, and there'll be no complaints." He lifted one arm to expose his watch. "Well, I'd better get on my horse now or they'll be sending out a search party." He turned a friendly smile on me. "Might even send that Toonerville character out, hey?"

I managed a small laugh and he contributed one of his own.

"He'd *find* you too," said Kidder.

"Oh, hell. He couldn't find his own way home. Of course he might make you his deputy." Curtis this time gave a full-throated laugh but Kidder could manage only a tight smile.

"No chance," he said.

"Good man," said Curtis and hastened out the door. "Jerry!" he called. "Let's go!" He stood in the cold and fished his stocking cap out of his coat pocket, then arrayed himself in the mighty coat, as if he were preparing for a descent into the sea. Gordon came trotting out from the kitchen.

"You're going back?" He was sincerely pained. "I thought you'd stay and have supper."

Curtis patted Gordon's arm, then took him again by both elbows and gave him a gentle shake.

"No. I have to get back. *Somebody* in this family has to work, you know!" His laugh brought echoes from the ridge behind us.

"Then you don't mind if I stay on? I've only got a few more days. Okay?"

His father was stumped by this for two or three seconds.

"Well, yes," he said. "As long as you're enjoying yourself." Apparently he was not entirely happy with the way

274

that sounded. "As long as you're not in the way. And don't do anything foolish. Jim knows how I feel about that." He questioned Jim with a quick, wide-eyed stare but Jim did not respond. Anger seemed to have struck him permanently solemn. But Curtis looked quickly away and hoisted the skirts of his coat so he could climb unimpeded into the pung. The horse flung his head impatiently, jangling all the bells. Jerry shook out the reins and shouted to the horse. The outfit moved swiftly away. Despite the cold we all stood watching it out of sight. Gordon's father half turned as they reached the top of the knoll and raised one hand high in salute. Only Gordon responded. Then we all hurried back to the warmth of the stove and the comfort of fresh coffee. It was too late, Jim decided, to try to get to the Beasley camp and back and there were still rooftops to be cleared closer by. We all sat silent again, with Jenny and Gordon both probably trying to think of some proper way to discover what had led to all the shouting. Kidder, however, concentrated deeply on his coffee and I carefully avoided meeting Jenny's eye. Finally Jenny could hold back no longer and simply asked the flat question:

"What was that yelling all about up there? He come to lay you out or something?"

Kidder had had a long time to plan his reply. "Oh, someone seen Bill Jones coming down with his pack, saying he stopped by and seen you folks, and the boss somehow got it into his head old Bill was using this as a way station for his booze run. He thought surer than hell the camp was going to get raided. He was ready to smell it out himself, but I managed to convince him Bill was in the dry goods business now. He still didn't like it too damn well. But as long as the Feds ain't coming . . ."

Jenny grinned in sudden relief.

"Oh, my land!" she exclaimed. "I thought we'd got you

in trouble, having us in here. It did seem to me for a minute like he warn't too pleased to find these extry people around."

Kidder waved the notion off.

"Hell, no. Just so long as you ain't running any rum."

"Well, he did seem like a nice man. Real nice." Jenny smiled at Gordon as she said this and Gordon shrugged.

"He's all right. He has a pretty short fuse sometimes. But he gets over it."

Kidder and I managed to shovel off one roof and replace a shanty cap before suppertime. The storm Jerry had forecast blew by without shedding a single flake and the weather began to moderate before the sun went down. Next day was mild and clear, with no wind at all, and the air sweet as fresh apples. Standing on the porch of the lodge, inhaling deeply as they had taught me to do in school, I thought I could smell the summer coming. But Kidder had no time for rejoicing at the weather.

"We better hotfoot it up the pond this morning and get a line on that Dunham lady, or whoever the hell she is. We don't want her down here stalking old Jenny if we can help it. We can make like we're headed for Beasley's. But you know one thing. If it *is* Kitty Dunham, we're going to have to tell Jenny. Wouldn't be right not to."

In the morning, to add conviction to our tale, we had Jenny put us up a lunch that we would, supposedly, consume at Beasley's, using the heater stove there to heat water for tea. When we had moved a good way up the lake, Kidder chuckled about this.

"Goddamn stovepipe's most likely plugged solid with mouse nest. We may have to eat our dinner at Kitty's."

We found no snowshoe tracks nearer to us than what we had seen the day before. But when we neared the cabin we could see the new tracks coming from the north. At the same instant, we both smelled smoke. I felt a tremor of fear in my gut. Kidder stopped and I pulled up at his side. He gently plunged his shovel into the snow and I set mine beside it.

"She's there," he said in a low voice. "Now what do we do?"

I wanted to turn and go back but I dared not say so.

"It's up to you," I whispered. "Only we don't know for sure if it's her."

Kidder thought about this for a second or two.

"Only one way to make sure," he muttered and plodded on up to the door. He had to sidle up to the steps to keep his snowshoes from interfering. He knocked rather gently on the door. There was no sound from inside. Kidder then pounded with his mittened hand, shaking the whole door. This time there was the distinct sound of a human voice inside. But whatever it said, it did not resemble "Come in." Nevertheless Kidder pushed open the door and stood for a moment staring. The voice inside this time clearly said "Who's that?" although the words were throaty and slurred together. I moved up close behind Kidder and shared his view of the scene inside. A gray-haired woman sat in a straight chair. She held a rifle loosely in her lap. Her mouth hung open, a hank of frazzled hair hung down across one eye. The room was warm, and wood still snapped in the stove.

"What you *want?*" the woman shouted, as if she had asked the question several times before. It all came out as one word, sounding more like "Wasshuant!"

"You all right?" said Kidder foolishly. The woman was far from all right. She was clearly too drunk to stand. But she did stand, all the same, using the rifle as a crutch to heave herself to her feet, then holding to the chair back to keep herself unsteadily erect. Her gaze, baleful, dull, and slightly wandering, fixed itself on me.

"You son of a bitch!" she snarled. "What you do with my brother's car?" Her words still ran together but the meaning was distinct. "You stole it, you bastard!"

This insane accusation almost struck me dumb. I mumbled, "Not me," or some such stupid riposte, then exchanged a look of half-amused bewilderment with Kidder.

"She's hotter'n a stove lid," said Kidder. "I'd like to get that goddamn rifle away."

"Jesus! Don't try!" I made an automatic step backward and almost went down, as my snowshoes got tangled.

"Don't worry. Anyway she's as likely to shoot herself as anybody."

"Goddamn you!" Kitty screamed, her voice now straining her throat. "I know you, you bastard!" She began to topple, and had to drop the rifle and grab the chair with both hands. Kidder cringed when the rifle went down, then began to make his retreat. I too, lifting my snowshoes with extreme care, worked my way into reverse. Kitty's screams continued to assail us. "You bring that car back, you bastard! You'll be goddamn sorry!" Kidder, half turned around, reached one long arm out, and yanked the door tight shut, muffling Kitty's screams only slightly. I had already started on the path.

"Let's get out of here!" I gasped.

"Oh, she'll never make it to the door." Kidder was actually laughing. "At least she'll stay put for a few days. Can you imagine her trying to hitch up her snowshoes?"

But we fled along our own tracks as if she were trotting right behind. Kidder called a halt when we reached the lakeshore. We were both blowing hard.

"She scares the hell out of me," I told him.

Kidder laughed again.

"Scared? Jesus! When I seen that drunken old bag with the thirty-thirty in her lap, I didn't know whether to shit or go blind! What's this stuff about her brother's car? I didn't know she had a brother."

"I don't know anything about her goddamn brother or his car. I can't even drive, for Christ's sake!"

"Well, you better steer clear of her until the liquor runs out. She still had two fifths of Peter Dawson ain't even been touched. And maybe they's more I couldn't see. She ain't going nowhere for a couple of days anyway. So what I'm going to do, I'm going to hightail it right down to Oquossoc and fetch old Stapleton. He'll haul her off to Augusty, which is probably where she should have been all along. When she's like this, they're obliged to take her. A nice long stretch in the bughouse will sweeten her up."

Kidder took the lead down the pond but I had taken only two or three steps when a memory jolted me.

"We forgot the shovels!" I called out. Kidder stopped and looked back.

"Jesus, yes!" He swatted the air with both hands. "Well, I ain't going back. She might have made it to the door. Not that she could even see straight, let alone shoot. Course she might aim at you and hit me! The shovels can wait."

"You going to tell Jenny?"

Kidder pondered that for half a minute.

"No. Let's not. No need for her to get her bowels in an uproar just yet. That old bag ain't going nowhere for a while."

"Maybe Stapleton'll shoot her."

"By Jesus, he will if he has to. Wouldn't hesitate a minute!"

By the time we reached the club, Kidder had his excuse all worked out so that when he delivered it to Jenny I found myself half believing it.

"We had to get off that roof in a hurry," he told her. "Them rafters was about to go. Another two, three inches of snow and she'll be flat. I'm going to have to prop that right off, if I can find the things to do with. May have to go down to the storehouse to get them."

Jenny showed no surprise, nor did she ask just what were the "things" that could be got only at the storehouse.

"You'll want dinner before you go?"

"*If* I go," said Kidder. But of course he was bound to go and he soon lost patience with the sort of make-believe search he put on, pawing over tools in the shed, and within ten minutes he came back to the kitchen to announce that he had grained the horse and was ready to start.

It was a fairly silent meal, for neither Kidder nor I dared speak, it seemed, lest we let a word escape that might ignite a whole string of questions. Gordon told us of his difficulties in getting the wire off a hay bale to provide the horse his breakfast and Kidder laughed a little.

"Take the axe to it!" he said.

Curtis wrinkled his brow.

"Is that how you do it? You cut the wire with an axe?"

"That's what the old axe is for."

"I sort of tried to untwist it."

Jenny, getting up to fetch the coffeepot, laughed merrily. She patted Gordon foldly on the shoulder.

"You could have spent all day on that job."

"Well, I pulled enough out anyway." Gordon gave me a sheepish half smile and devoted himself to his venison.

Kidder was on his feet in minutes and took two gulps of his coffee before he put his coat on. He seemed slightly dismayed, as I was, to note the package of lunch bulging from the game pocket of his coat. He grimaced at me, glanced quickly at Jenny, and hurried to the door.

I gave Jim a hand in hitching Billy to the sled, then Gordon came to stand, coatless, by my side as Kidder climbed into the seat. Kidder looked into my eyes for half a minute, as if he wanted to warn me of something, but merely cried out "See you later!" and urged Billy on with a slap of the reins.

With Kidder gone I felt suddenly as if I had come out in the cold without my shirt on. My God! There was nobody here but just Jenny and Gordon and me! I looked around at Gordon, who was on his way to the kitchen hugging his ribs. Then I looked up the glittering lake. Jesus! That was one hell of a long lonesome stretch! And up there beyond the tree line there was that crazy, drunken old lady with a loaded gun. I supposed the right thing to do was to hike up to our cabin and get Kidder's rifle out. And what the hell would I say to Jenny? I knew that what was on Kidder's mind when he left was that I should keep my mouth shut about Kitty. Unless of course the crazy old bitch should come charging down. . . . Oh Christ, I told myself, she can't even stand, let alone walk. As for putting on her snowshoes . . . This thought brought me real comfort. Even sobering up and shaky, she'd have one hell of a time dealing with those frozen straps. So I looked up toward that distant, unseen little hideout and wished with all my heart that she would open another bottle and drink until she fell on the floor. Then I turned back to the kitchen and joined Gordon and Jenny in drinking lukewarm coffee.

I sat there silent, listening while Gordon told Jenny of the hazing he had undergone when he joined his college fraternity — the paddling that had raised blood blisters on his "behind," and being made to carry a chamber pot full of cider down into the center of town, then publicly drinking the cider and eating the cruller that had been floating in it. I laughed dutifully but I was uneasy all the same lest Gordon turn to me and invite me to tell a few college tales of my own. It was a matter of shame to me, at that age, that, in my brief stay at college, I had not been invited to join *any* fraternity — indeed, had been emphatically disinvited — and if Gordon had asked me I probably would have lied about it. Instead I tried once or twice to steer the talk into a new course only to be dismayed to find myself about to tell of what we had found away up there in the snows, this side of the border. So I got up finally, offering the excuse that I had to fetch fresh kindling and firewood to the cabin. When I had finished this chore, I found that I had run out of jobs. There were roofs to shovel but my shovel and Kidder's were both standing forlornly in the snow in sight of Kitty Dunham's retreat and, even though my worry about Kitty had subsided, I was damned if I would hike up there within range of her wild eye. She could hit me from the doorway.

The proper thing for me to do of course was to chug my way back to the Beasley camp, where, as far as Jenny knew, we had left our shovels. But with twilight coming at four o'clock that would clearly make no sense at all. So, after killing a little time watching Gordon set new distance marks with his skis and refusing again his earnest offer to try the sport, I took myself to the kitchen porch and ostentatiously "borrowed" the heavy wooden shovel Gordon had been using to keep the pathways clear. With this on my shoulder and snowshoes bound to my boots I set off to the Corcoran camp — a long log building on the western shore, set on posts high enough to be almost clear of the drifts, where a long porch roof and an extra guest camp, both solidly blanketed in snow, offered work aplenty to keep two men off the ground for a full day or longer. I managed to complete about half the porch roof before the sun was gone and I could decently knock off and start back for supper.

Once or twice on the way back, I looked behind me up the long empty lake and took comfort in the distance between me and the beginnings of the forest. No one could approach across that barren stretch without being in view for twenty minutes or more — time enough to make ready, one way or another.

Gordon, come down to the lake to fetch water for the kitchen, awaited me by the water hole. I took my snowshoes off there and carried one of the pails, leaving my snowshoes stuck tailfirst in the snow by the water hole. A pleasant weariness took over me and a mild ache tingled in my shoulders.

"What's for supper?" said I.

Gordon laughed.

"Venison. For a change."

"Can't make me mad."

Jenny turned from the stove to greet us, her smiling face all a-flush from the fire.

"Almost ready," she said. "You still got time to wash your forward feet."

This was a standard joke with Jenny and we both rewarded her with the standard chuckle.

The warm kitchen and the warm supper seemed to quiet the last flutterings of fear in my heart. What could happen now? It was dark. Kitty most likely lay unconscious on her bed — or on her floor. Jim Kidder would be back in the morning. As I told myself these things I realized suddenly that I had been staring right into Jenny's face all the while. She smiled at me quickly, as she had smiled at me a half a hundred times. She seemed to have read my last thought.

"Sure be glad to see that old Kidder," she said. "Seems like he's been gone a week."

And there sure will be hell to pay when he gets here, I told myself. I found I actually anticipated the excitement that would follow when Stapleton set out to consign Kitty to the state asylum. Would they come with a straitjacket?

Would they tie her to a sled? These questions nearly burst out of me as I tucked into my tea and cookies.

"What's the joke?" said Jenny, who had never stopped watching my face. I was surprised to learn I had been smiling.

"Nothing much. Just something Kidder said once. He sure has a funny way of putting things."

I filled my mouth with all that was left of my tea and picked up a cookie to take with me. It took two gulps to down my tea.

"I'm pooped," I announced. "I'm hitting the hay."

Jenny looked really dismayed.

"It ain't even eight o'clock."

"I know. But I'm falling asleep already." I scooped up my hat and jacket and made off before anyone could ask me what it was that Kidder said that had made me smile.

My cabin was cold. When I had the lamp lighted I set about laying birch bark and kindling in the heater stove. The crumbling remains of the deer head stared up at me and when I poked at them to make room, sparks drifted about. I stirred the embers then and needed no match to start the birch bark flaming. It was a full ten minutes before I had fire enough to warm me. I stood there offering my hands to the heat and wondered if I would ever get to sleep, from fretting over Kitty Dunham. But when the warmth finally found me, I shut the damper on the stovepipe, shed my outer clothes quickly, doused the light, and, as Kidder used to say, tossed and turned for the better part of a minute before sinking into a stupor.

I woke up when my window was gray and lay still for a moment wondering why I felt so glad. Then it came to me of course that Kidder was due and that it would be too late, when he arrived, for Kitty to wreak any harm. I hopped out barefooted to feed the fire, then crawled back into my blankets to let the cabin warm up. Should I go up with Kidder and Stapleton, I wondered, to lend a hand in the capture of Kitty Dunham? There came a dreadful emptiness in my stomach at the thought. Shouldn't someone stay with Jenny? But when I tried to imagine myself putting that argument to Kidder, I knew I would never utter it. It would be just too damned unmanly. So I envisioned myself facing that madwoman once more, assuming a brave face, perhaps even helping to manhandle her into submission. I went to breakfast with little appetite. But while I dawdled over my coffee, I was inspired to believe that if I were dutifully engaged in clearing the Corcoran camp roof when the expedition took off for Kitty's hideaway I might expect to be left behind. So, despite my urge to be on hand when the rescue force arrived and to share in the uproar that would follow Kidder's announcement of the true purpose of his journey, I took up my shovel, took my leave, took care to remind Jenny where I could be found, and took myself off for Corcoran's.

I did not fail, as I plodded up the lake, to pay close heed to what might be afoot at the distant northwest landing. But there was no movement there, nor anywhere on the whole white expanse to the east and west and north. It was a bright cold day with no wind and no birds flying. By approaching the Corcoran camp from the south I could unhitch my snowshoes and with a minimum of wallowing hoist myself to the porch roof, shovel and all. There were ten square feet of two-feet-deep snow to shovel yet. But under the crust it was nearly as dry and powdery as baking soda and it was no great chore for me to lift whole squares of it at once and fling it off the roof. It took the sun, edging out of the woods to the south, more than half an hour to find me, although it had set the northwest end of the lake aglitter an hour earlier. After a few minutes I forgot to listen for harness bells and lost myself in relishing the joy and relief that would follow when crazy Kitty had been packed off to distant Augusta. My soul then was perfectly serene and I was trying to work another whole square of snow onto my shovel when there came a splintering crash as the small window I was facing shattered in my face. Without thought I leapt backward, the shovel still in my hand, and went tumbling off into a deep drift where I immediately sank the width of my body. It was only when I lay there, with nothing but empty sky above me, that I heard the rocketing echoes that seemed to cascade from every mountain.

"Oh, my Christ!" I gasped. For that had been a rifle shot. I actually began to weep with terror, thinking of how far off any help might be. He should have come *sooner!* I wailed inwardly. You shouldn't leave me here! I dared not move, not even breathe, for uncounted seconds. I was sure I could hear pounding footsteps coming close. Then I realized it was my own terrified heart racing to give me the strength to run. The shot, I knew, had to have come from the woods north of the camp, for with every shovelful I handled I had kept watch on the lake. No one had come down in the open. At least for now, with a little scrambling, I could get the camp between me and Kitty and could reach my snowshoes.

Desperately floundering a foot or so at a time, in snow more than hip-deep, I reached my snowshoes and made a sort of sled of them, on which I could creep into the woods. Among the trees I stopped for a while to catch my breath. I had lost one mitten along the way. The snow had not drifted here and I was able, after much toppling and straining, to shove my feet into the snowshoe clips and to get both the straps fastened. I crouched then and listened for two or three seconds. There was no sound from beyond the cabin. At least I would have some sort of start, for if she had been near I would have heard the snowshoes crunching. With my head bent far forward and both hands catching hold of branches for traction and support, I made my way, wildly, through the evergreen tops toward the camp. I thought of shouting for help. But that would only inform Kitty I was alive and moving. Perhaps she had counted me dead when she saw me plunge into the snow. Perhaps she feared the shot might bring help and she was already fleeing.

I had never known before that it was possible to run on snowshoes. But I ran. And I tripped and fell no more than four or five times before I burst into a clearing and saw, down on the lake, Jenny standing alone. She was obviously looking for the source of the shot, thinking perhaps it had been a sort of signal, or Kidder doing a bit of poaching on the way home. I dared not shout, lest I give Kitty something to target on but I could not let Jenny stand out there in full view. I waved my mittened hand frantically in the air, moving down toward the lakeshore, where she might see me more easily. She caught sight of me as soon as she had turned my way.

"You hear that?" she yelled. "Where was it?"

I could not let her stand there and invite slaughter. I had but little breath to spare but I managed to yell:

"Get out! In the woods! Woods!" And I beckoned her fiercely with the whole length of my arm. She smiled at me and came toward me at her regular pace, yelling some sort of question in return. But I paid no attention. I came a few strides closer.

"Quick! In the woods! Somebody's shooting at us! Hurry!"

Rather reluctantly, and still smiling easily, she hustled up to me. I turned and beckoned her to follow. She did not hurry.

"Just some poacher, most likely!" she called to me. "He'll be hightailing out of here soon as he hears our voices."

"No poacher, for Christ's sake! Someone tried to shoot me!"

Jenny looked concerned but obviously not convinced. She came close to me, frowning and shaking her head.

"You mean he *tried* to hit you? How do you know?"

"I'll tell you later. Come on!"

It is strange how much more secure I felt with someone at my side. But fear still urged me to hurry. Jenny had trouble staying near.

"Wait up a second, can't you?" she panted. "Tell me what happened."

I barely turned my head.

"When we get there!"

In the camp clearing at last, I stopped, with my chest heaving painfully, and looked back up the lake. There was no movement anywhere. The thumping of my heart filled my ears. When Jenny caught up to me I still lacked the breath to talk and I simply motioned toward my cabin. She followed me, slowly now, and I had my snowshoes off when she reached the steps.

"Where's Gordon?" I gasped.

"Hell, I don't know. Off on his goddamn skis somewhere. Don't tell me *he* shot at you."

I went into the cabin and took Kidder's carbine out of the corner. There were shells on a ledge over his bed and I began to slip them into the magazine.

"What the *hell* is going on?" Jenny's voice was nearly a scream. I plopped down on the bed and took the time to get more breath.

"Kitty Dunham. We should have told you before. She's been staying up at that shack on the border. Kidder went to get Stapleton. She's crazy. . . . They'll . . ."

Jenny did not let me finish. Her face seemed drained of blood.

"Kitty DUNHAM! It was Kitty DUNHAM!" The blood came back to Jenny's face in a surge. "Did that crazy bitch take a shot at you?"

"She sure as hell did. She thinks I stole her brother's car, for Christ's sake. Kidder and I were up there and she was falling-down drunk. She must have . . . well . . . she came down through the woods. We better stay here until Kidder comes."

Jenny hadn't been listening.

"That dirty old whore! She *shot* at you!" She kept looking me deep in the eyes. "She better not show her goddamn face around here! She *shot* at you? Goddamn her dirty soul!" She went to look out the window, up the lake. "You stay here and keep watch. I'm going to fetch my *own* gun. I'll blast that old bitch to kingdom come!"

I had never before seen Jenny so fierce, nor so wild-eyed. Her nostrils flared as if she were leading a charge up the hill.

"You stay here," she said again and she hurried off toward the kitchen, carrying her snowshoes. I closed the door and took a stand by the window. I could see all the lake except the extreme northwest corner. And I could see the whole stretch of woods that bordered the dooryard. Kitty could not come close without my spotting her. But I had begun to believe now that she had reckoned me dead and had gone back to her camp, most likely to break open that last fifth of Peter Dawson. After a long wait I thought I saw a movement where the brook made an opening in the woods, to the east, just below the camps. I watched for a long time but saw nothing more. There was a tote road there where they brought in the winter wood. It was most likely Gordon, coming back on his skis.

Come *on*, Kidder! I whispered to myself. It was now past nine o'clock and Kidder must surely have been on the road by seven. He was overdue. *Long* overdue. Jesus! Could Kitty have waylaid . . . I refused even to think the thought. Kidder was too damn smart for that!

It was just then that I heard the crunch of feet on the crust in the yard. I dodged out of sight, listened for a moment, then peered very cautiously out of the window, through the opening between the dingy curtain and the window frame. I put my rifle at half cock. But it was only Gordon, standing on the slope to look down the lake. He stood quite still for a moment, then came crunching on toward my cabin. I opened the door to greet him.

"Where's Jenny?" he called. I felt a sudden chill in my gut. Holy Jesus!

"Isn't she in her room?"

Gordon shook his head. He seemed to take fright from my expression. His voice dropped to a half whisper.

"Something wrong?"

"Christ," I said, "I don't know . . ."

"I heard that shot a while back."

I stared at Gordon for several seconds before I decided to tell him what had happened.

"There's this crazy old woman up at the border. She took a shot at me. We think it's the same one took a shot at Jenny. Jesus. We've got to find Jenny. Quick."

Gordon stood, like a slightly stupefied child, his mouth half open, looking at me in disbelief.

"Goddamn it," he whimpered, "I *knew* we should have got Dad into this."

I felt a sudden hot stab of anger and had to swallow the first words that came to me.

"He can't do a damn thing for us now." I tried to keep my voice level but I was close to crying myself. I had to hold my mouth tight shut for a moment while I swallowed the sob that rose into my throat. "Was there any sign of her?"

"There was some fresh-looking tracks down below here. I thought they were coming here, but they slanted down over the pond and into the woods."

"Let's go," said I. I picked my rifle off the chair behind me, keeping it at half cock, and held it at my side as I hitched my snowshoes. Gordon had turned very pale indeed and he stared at the rifle as if it might coil up and strike.

"My God," he whispered. "You don't suppose . . ."

I was really too scared to suppose anything.

"I didn't hear any shot," I said, trying hard to keep my voice deep and steady. "I'll go along her track, if I can find it. She's not in the shanty?"

"No. I was in there."

"Maybe she went off to the gate to get help." I was trying to convince myself. I had no idea how far it was to the gate. I would just follow her track and hope to find her. If that *was* her track. I smothered a dark suspicion that she had simply run off to hide somewhere. Oh God! Why didn't Kidder come!

Gordon was thinking the same thing.

"When's Kidder coming?"

"He's due any minute. You stay here and tell him what's happened. He'll have Stapleton with him and they'll know what to do."

"God. Stapleton!" Gordon whispered. "Listen, shouldn't I come along? It doesn't seem too *safe* here. Couldn't you wait until they get here? If they're due any minute?"

"I want to find Jenny!" This time my voice really broke, so that I sounded like a boy who had cried "I want to find my mother!" Oh, Christ! If anything had happened to her. To get my shame out of his sight, I started at once down the slope.

"I'm going to get into the kitchen," Gordon called. "I'm staying in there."

I made no answer. I had already seen Jenny's tracks. She seemed to have headed back along the trail we had made coming in, but she had kept out of the woods and had gone up a road I did not know existed that seemed to run along behind the private camps. What the hell am I going to do? I asked myself. I don't recall ever having been more scared in all my life. I was so goddamn *alone!* My knees seemed always about to buckle. My stomach had gone hollow. I had to bite my lips to keep from whimpering. I stopped two or three times and silently prayed to God that I would hear those harness bells behind me. But I kept going, hoping I would see Jenny around the very next turn. Her empty tracks went on and on. At least she was alive just minutes ago. And if I hurried . . . If she was just trying to get away, I could go . . . Only we couldn't leave Gordon! No, I comforted myself, Kidder will be there sometime. I stopped dead still then to ask myself if it wouldn't be better after all to go back and wait for Kidder. Jenny might be so far ahead now that I would never catch her. And I would not want Kidder to think that I too . . .

At that very moment, I heard the rifle shot. It seemed far ahead of me, at the end of the lake. The echoes came back endlessly from the border mountains. Oh, Jesus! My first impulse was to turn and flee. But some sudden streak of courage, or perhaps desperation, seemed to brace my spine. No, by Jesus, I would hide out here. If the old bitch had shot Jenny I would waylay her right here. I could see some fifty yards ahead on the road and there was brush enough here to conceal me completely. The moment she showed her ugly face around that turn, I could pick her off. I lifted the rifle and sighted down the barrel, choosing a tree to hold the sight on. My hands were steady. I knew I could hit her at that distance. The coppery taste in my mouth was the same I used to feel at school when I waited in the corner of the ring for the bell to start the bout.

It was at least twenty minutes before I heard the soft sound of snowshoes on the crust. Instantly I forgot the cold in my toes and hands and I went to one knee on top of my snowshoe, drawing a bead on the empty air at the turn in the road. My pulse beat in my ears like a swiftly tapping finger. The coppery taste returned to my mouth and my breath whistled faintly in my nose. There! There was movement in the brush. Then the figure appeared. I adjusted my gun and sighted. . . . Then I cried aloud. It was Jenny! I jumped out of my ambush and shuffled toward her, laughing.

"Jesus!" I yelled. "I almost shot you!"

Jenny looked at me without any joy. Her face was drawn and white, her eyes dark. I stood still and let her approach. She shook her head at me, like someone who had bad news to impart.

"What happened?" I said.

"I shot the dirty old bitch! I winged her!" Jenny bit her lip as she waited for my reaction.

"You didn't *kill* her!" I was trembling with excitement and my voice cracked.

"No. I hit her in the leg. I guess. Anyway she went down. Right by her cabin. I stayed there until I saw her moving. She made it in the door all right, I guess. Holy Jesus! I don't know what to do! I almost wish I never . . . But goddamn the old whore, she tried to *kill* you. You're not supposed to just sit around and let someone take potshots at you."

"For Christ's sake, no! You did right, Jenny. It's okay. You *had* to stop her."

Jenny kept nodding as she accepted my comfort but she did not cheer up.

"I know. I know," she kept saying. "But Christ, what a mess! Surer than hell, they'll pinch me. They *have* to if you shoot somebody. Why couldn't the old bitch just have gone back to her bottle?" Jenny lifted her agonized face to me. Tears covered her face, wetting her chin and dripping on her jacket. "Oh, why the hell did this have to happen? Why couldn't we just . . ." She took hold of my arms and laid her head against them. Our snowshoes kept us from getting close.

"Well, come on home," said I. "Kidder will figure out something."

"I still don't feel right about leaving her. One of them shells can make an awful hole."

"Stapleton and Kidder will be here soon. They'll tend to her. Jesus, she probably would have shot you if you came close."

"She dropped her rifle. Jesus, I just feel awful. You think we'd ought to go back and see . . ."

"And get shot? She's got her rifle back by now. Or she may have a pistol. Christ, Jenny. Don't be nuts! You can't go back there! Leave that stuff to Kidder."

She nodded glumly and we started back along our tracks. It was easy walking on the crust, but we did not hurry. We had just about reached the rear of the Corcoran camp when we heard the harness bells.

"That's them!" I cried and I turned at once to the lakeshore, to hail them. But Jenny grabbed at my arm.

"No! No! I ain't ready to face Stapleton yet! Give me time to get my story ready. Jesus, I don't know what I'm going to say."

"Just say what happened."

"That I went gunning for her? How's that going to sound? How does anybody know she shot at you? I can just imagine the story Kendall and them can cook up. Let's get home and think about it."

As we stood there, the sled came into view, old Billy pulling it up the flat frozen surface with only a faint jangling from the bells, gently, steadily, remorselessly — like a gun crew, I thought, going into battle. There were three men on the seat, snuggled tightly together.

"Who's the other guy?" I whispered.

"Doc Morton. You can tell him by that Christly red hat. God, I'm glad they got a doctor."

"He's in for a surprise."

"Oh, they've probably got him scared shitless already." Jenny managed a very slight snicker.

Back in the kitchen, we must have sat for more than an hour sipping coffee and nibbling molasses cookies, and coping with the questions Gordon asked. Jenny, who would take a fit of trembling now and then, as if she were sitting in a cold draft, would not tell Gordon she had shot Kitty.

"I chased her back where she belongs," she explained and looked solemnly at me to make sure I was not going to tell any more of the story than that.

"And she didn't even shoot?"

"She threw her gun away."

"God!" Gordon fixed Jenny with the gaze of an adoring small boy. "I swear, you've got guts! I swear, there are plenty of *men* who wouldn't have dared! . . . I mean, that took real *guts!*"

Jenny was too distraught to render the traditional shy disclaimers. She would squirm in her seat from time to time and take her lower lip between her teeth. I knew, almost as if she had spoken aloud, that she was asking herself what in God's name she was going to say to Stapleton. But we had no time to talk about that before we heard the harness bells. We all got up to look out on the lake and saw that the sled was much closer than it had sounded. It did not come up toward the kitchen, however, but swung straight up the slope, where they must have broken out a sort of road on the way down. Promptly the whole rig — horse, sled, and all its burden — disappeared behind the drifts. My impulse was to go to meet them, but Jenny held me back. I knew she needed more time. Gordon, however, rushed out the door, leaving it open behind him, and raced up the shoveled path as far as he could reach without wading into the drifts. Jenny pulled me back.

"Sit down a minute," she whispered. "What in Christ's name am I going to say?"

"You can't make up a story. Just answer his questions. Tell him what really happened."

But apparently Jenny was really just asking the question of herself. She did not seem to hear what I told her and her chin was atremble as if she were ready to start crying again. Then we heard the footsteps coming and Jenny sat down and stared wide-eyed at the door. Kidder came in first and we could see Stapleton, snowshoes in hand, right behind him on the porch.

"Well, we took the critter," Kidder said. He seemed still out of breath from a struggle. "She come peaceful as you pleased. She got herself shot in the ass somehow and she'd been bleeding some. Had on a pair of old Jack's britches, I guess, and they must have soaked up a couple of quarts. Talk about your mess." Stapleton came in, crowding past Kidder to look for the coffeepot and right behind him came Doc Morton, holding his famous hat in his hand. He was a rotund little man, with just a few strands of pale brown hair plastered to the top of his head. He took off his tortoiseshell glasses and squinted at us.

"Oh, coffee!" he said. "Good!" He took a clean cup from

the cupboard as if he were at home and stood behind Stapleton, waiting his turn.

"How's the patient?" said Kidder. Morton turned around and made a slight snorting noise that might have been a laugh.

"Oh, resting easy! Resting easy! I gave her enough to hold her for a couple hours."

The men all took chairs at the table and passed the sugar bowl about. Morton kept shaking his head.

"My! My! That *was* a trial! I could do without that sort of thing!"

"Well, you *will* make house calls!" said Kidder.

Morton gave a mirthless laugh. "Yes. Well . . ." He looked straight at Jenny. "That young man out there said you chased the lady back to her cabin? You see how she get shot?"

"Couldn't make out," Jenny mumbled. "I seen her on the snow."

"Well, it was just a flesh wound. Bled like sixty though. Of course she had plenty of flesh to spare." He looked all about to see his joke take hold and he laughed again, more heartily.

Stapleton had been studying Jenny all the while. His face was relaxed, nearly expressionless, as always, but I thought there had been a slight twitch at the corner of his mouth.

"You heared the shot though," he declared quietly.

Jenny merely nodded. Stapleton nodded back at her.

"Some poacher," he said. "Had to be. She couldn't have shot herself that way. Warn't no powder burns anyhow. One shot missing from the magazine." He turned his cold clear gaze on me. "That was the one meant for you, I suppose. Young feller told us she tried to pick you off."

I shifted and coughed; I had not been prepared for any examination.

"It came goddamn close," I said.

"I don't believe she was shooting too straight," said Stapleton, smiling now. "Not from what we see up to the camp there."

The doctor gulped the rest of his coffee and stood up.

"Well, we can't leave the poor soul to freeze to death."

"Got her bundled up good," said Kidder.

"Yes, but being immobilized that way. And losing that blood . . ."

Stapleton was the last to stand up. He had not taken his eyes from Jenny.

"You look pretty peaked," he said. "You'd ought to come back home with us and get rested. They been running you pretty ragged."

"May be," said Jenny. The tears were already flowing on her face.

"Go ahead," said Kidder. "I can keep this crew fed. You done your share."

"She did more than her share," Gordon announced in a shrill voice. "She has a lot of guts!"

"I always knowed that," said Stapleton. "Come on now. Get your gear together and we'll give you a ride down."

"Yes, yes," said Morton. "We really should have a woman along. It doesn't look quite right, really, without a woman."

"All right," Jenny whispered. She dodged into her bedroom and spent ten minutes stuffing her clothing into a bag. The doctor went out to the sled but Stapleton stood waiting in the kitchen. He looked from Kidder to me.

"*Had* to be a stray shot," he said. "That the way you figure it, Kidder?"

"Had to be," said Kidder. Stapleton grinned suddenly at me.

"Anybody trying to kill her never would have shot her in the ass! Right, young feller?"

"Absolutely," said I.

"Well, they's been poachers around," said Kidder. He too grinned at me. "We seen the signs. Hey, Bob?"

"Right."

Jenny came out with her white bag all stuffed. I noticed that she had not brought her rifle.

"I left a few things," she murmured to me. "You look after them."

"Sure."

She took me suddenly in her arms and kissed me. They all went out then and I watched them out of sight beyond the drifts. Then I went into Jenny's bedroom and picked up the rifle to take up to my cabin. I heard the harness bells jangling. In a few minutes, Kidder came back alone.

"Our boy's gone down with them," he said. "He's going to bring the team back."

"Think you can trust him?"

"Well, Billy knows the way."

But whether Billy knew the way or not, he did not return until almost twilight, when Kidder had begun to fret that he might have to go out and guide our lost boy home with a lantern. We had not quite abandoned hope when a faint tinkle of the harness bells sounded from beyond the knoll. Kidder and I were both standing on the road when the sled came up the slope. But Gordon was not driving. A bulky figure in a stocking cap hunched in the driver's place, clucking earnestly to the horse (who needed no urging), and holding the reins chest-high, as if he were handling a coach-and-four.

"Holy jumped-up Christ," Kidder breathed. "That's the old man again. Must have another bug up his ass."

The elder Curtis, however, was full of good cheer when he pulled the sled up beside us, with much hauling on the reins and shouts of "Whoa, there! Whoa!" Apparently he had no zest for taking Billy to the hovel, for as soon as they came to a halt, he threw the reins up over the horse's back and set out to alight.

"Greetings and salutations!" he cried. He lifted the skirts of his regal coat knee-high and jumped to the roadway.

"I didn't like the idea of Gordie's trying to make this alone," he told us. "This time of year. Dark comes on pretty fast."

Gordon let himself down rather reluctantly and growled half to himself: "For God's *sake!*"

"Get back there!" Kidder shouted. "Take your steed to the hovel! Hell of a teamster you are!"

Gordon climbed back in one jump and gathered up the reins.

"Get up," he cried, shaking out the reins like a veteran. He dealt his father a look that was anything but agreeable.

"Hope you fellows can make room for me," said Mr. Curtis. "I'd like to stay now until Gordie's ready to come out. If you think you can put up with an old geezer!"

"We'll manage, somehow," said Kidder. "You might have to take turns in the bunk with Bob. You could have the late shift."

Mr. Curtis looked truly alarmed at this and he turned a sharp glance upon me, as if he might order me to find other quarters. Kidder laughed. "Just joking. You can have a room to yourself in the main camp, next to Gordon."

"Well, now, don't fuss," said Curtis, his feathers all smoothed out again. "I've been in the woods before, you know." He conferred a pleasant smile on me and nodded after the vanished sled. "Be a good fellow, Bob, and fetch me my bag." I trotted after the sled and found Gordon tinkering doubtfully with the harness. "Leave it," I told him. "I'll do it. You take your old man's bag up to the camp." We were out of sight of the others, but Gordon sent a mildly disgusted stare in his father's direction.

"Well, thanks," he said. He lifted a large grip of glistening tan leather out of the sled and took off glumly for the camp.

When I returned to the kitchen, Mr. Curtis sat at the table with a green bottle of Peter Dawson standing before him. He held a tumbler nearly half full of amber liquid, while Kidder, busy at the stove, had a drink of his own on the sink beside him. Mr. Curtis smiled at me in his tight-mouthed way and gestured toward the bottle.

"A little of the cup that cheers?"

"I don't mind if I do."

I was not much given to drink in that day, for the taste still made me gag a little. But it was the part of a man, I knew, to learn to choke this stuff down and say it was good. So I dumped a double mouthful into a tumbler and lifted my glass in a silent toast. Curtis solemnly returned the gesture and took a large gulp from his glass. Gordon, who had been sitting silently at one side, watched me sip my drink, then turned to his father.

"Don't I get in on this?"

His father stared at him for several seconds, weighing the matter.

"Well, all right," he said. "But a light one now. A light one!"

Gordon trickled about an eighth of an inch into a tumbler and rolled his eyes at me. I was sure he had knocked back many a drink far stiffer than this. He toasted me with half a smile and did away with the drink in a swallow. His father, his chair turned partly away from us, was devoting his attention to Kidder.

"Another thing, Jim, that I'm mighty relieved about. I'm damn glad you've come to your senses about that other business."

Kidder gave the heavy spider a shake and turned toward the table.

"What's that?"

Curtis gave Gordon a swift glance over his shoulder. He turned back to Kidder.

"I mean getting rid of that White woman. You know that wasn't right, her being out here this way. She's a pleasant enough woman. But you know. Well, she's not *really* the sort you want . . . Well, you know what I mean."

"Nothing wrong with Jenny," said Kidder. "I was damn glad to have her."

Gordon got up before his father had a chance to reply.

"You want a chaser with that, Dad?" He pointed at the tumbler.

His father looked down at his glass and nodded vigorously.

"By God, yes!" he cried. "It's been so long since I've tasted that good old Winnebago springwater."

I opened my mouth to speak, for I had just seen Gordon carrying that bucketful up from the lake. But Gordon fixed me with a scowl that struck me silent. He took a fresh tumbler from the shelf and very ceremoniously dipped water out of the pail to fill the tumbler to the top. He handed it to his father and stood by while his father drank it down. Gordon and I both watched as his father drained the tumbler, his Adam's apple bobbing. Gordon seemed to be counting each swallow. When his father had finished, Gordon took back the tumbler.

"More?"

"No," his father said. "That was great! Delicious!"

Gordon turned to me then and held his eyes fixed on mine for what must have been the count of ten. His face was completely expressionless. The face of a conspirator.

❧ 10 ❧

WE had bright days then and bitter, cloudless nights when the surface of the snow turned to ice, so we could walk where we pleased, without snowshoes, slumping only rarely, as when we walked close to the trees where the top snow had not melted enough to freeze hard in the night.

Kidder and I spent less than a day cleaning the snow off the rest of the Corcoran camp and Kidder took only a half hour to find a piece of scrap wood he could fit into the window that Kitty Dunham's rifle shot had smashed. The camp then, with one eye blinking shut, took on a rakish look that made Kidder uneasy.

"I'll have to cut a piece of glass to fit in there before old Corcoran comes," he muttered, "or he'll be having kittens."

As we made our way across the glittering lake, toting our shovels, Kidder kept looking back, to get different angles on the Corcoran camp and mark how noticeable the blemish was.

"Christ," he complained to himself, "you can see the son of a bitch across the pond. I'll have to go down to the storehouse to get some glass, before the people start coming."

The people, however, were not going to be coming until late May and it was hardly the beginning of March now, so there seemed no urge to hurry. We had the MacLean camp to do now — the farthest removed, except for the Beasley camp, from the main lodge. It stood high above the water, with a long wooden pier that reached out above the ice for twenty yards. It was actually three camps — a large four-room camp with a kitchen, connected by a long porch and a short flight of steps to a smaller cabin on a slightly lower level, known as a mother-in-law's camp, then a smaller building in the rear called the guides' camp, where the lesser breeds were housed. All had windows snugly boarded and chimneys capped, secure in their Sabbath sleep from wildlife and weather. It was a neat-looking array, but Kidder was not happy with it.

"Look at the way them fucking fireplaces pull the corners down," he growled. "Next thing, one of them chimneys will be in the pond. Every time the frost comes out the fireplaces sink another few inches. The horse's ass that put this up just didn't feel like digging."

The long roofs seemed to carry a half ton of snow, crusted hard, nearly knee-deep, glinting like glass. It took us three whole days to finish, with long spells when Kidder and I had to go back to feed the Curtises and keep the elder Curtis content with long answers to his questions. He had heard talk of an Indian grave, somewhere on Brown Company land near the four-corner post that marked the meeting of four townships, two or three miles below the lake, and he seemed to be making ready to award Kidder the job of finding it. Kidder did not volunteer. And when we were alone he vowed that if the old son of a bitch insisted, he would get some rocks together and build him a goddamn grave as good as any in the country. Only there'd be no Indian in it.

In this time we managed to wipe out the forequarters of the deer and had even dug one hindquarter out of the snow and taken steaks off it. The elder Curtis consumed his allotment of this contraband with no comment except a slight raising of his brows when he first found it on his plate. He apparently had no appetite for another set-to with Kidder. All the same, Kidder, in preparation for the trip outside, waited until the Curtises were off to see how the family camp was wintering before he carried the rest of the carcass into the walk-in meat locker behind the kitchen, where it would be safe from the foxes.

I walked the crust over to the Curtis camp then to see if I might lend a hand at something. Their cabin was a sprawling log structure built on a knoll some hundred yards from the water. Here I found old man Curtis in a state very close to hysteria. Curtis, like all rich men to my eye, had always seemed neater and more closely shorn than others, not simply from his well-tended clothes, which were ostentatiously "rough," but from the very appearance of his flesh. But now he seemed, with his thin gray-brown hair awry, with several wet strands down over his brow, almost disheveled. His face and neck were gorged with blood and his eyes gleamed with moisture. His mouth was open when I came in, not in a shout but in a grimace of dismay and anger, so that he seemed about to weep.

"Look at this mess!" he charged me. "Look at this MESS!"

He flung one arm as if to indicate that the whole place was in disorder, but I could see hardly any mess at all, except a narrow mattress, obviously out of place, lying on the floor, with a large hole in it where some stuffing had been extracted by a mouse or squirrel. I was far more taken aback by the utter collapse of the man's smug self-

305

possession — the air that held him aloof even when he was being his most comradely. I looked at Gordon, who was standing transfixed a few feet away, glumly studying the offending mattress. He gave me one brief glance and offered the slightest movement of his head to suggest bewilderment.

"The thing was brand *new!*" his father shouted, bringing one closed fist down to strike his thigh. "Brand *new!* Just out of the *box!* What in the name of all that's holy got *into* that woman! How many times did I tell her to make sure that drawer was *tight!*" With this he dealt a kick to the slightly open compartment that protruded from beneath a built-in bedstead. "You must have heard me tell her a hundred times!"

I could only assume that the woman in question was Gordon's mother, for Gordon made no reply. His face had gone white and he faced his father with the grimmest of expressions.

"How many times?" his father demanded, although it did not seem any answer at all could mend the situation. Gordon swallowed hard and looked intently at the floor.

"I may have forgotten it myself." His voice was only a single decibel above a whisper.

This amazing confession merely seemed to pump his father's rage a little nearer to bursting.

"*You!* When in the name of GOD did I ever leave such a job to *you?*"

Gordon, speechless again, merely shook his head and resumed his silent contemplation of the wretched mattress. What prompted me to speak — unless it was an arrow of inspiration from on high — I cannot tell. It was such a simple thing to say, yet it rendered the whole scene too ludicrous for belief.

"That can be fixed," I said.

There was, for a moment or two, silence supreme. The elder Curtis turned on me, his eyes made doubly wide by indignation, as if I were a butler who had dared butt into a family quarrel. He glared at me and I looked innocently back at him, blushing, I am sure, but still too convinced of the purity of my intent to be even slightly daunted. Surprisingly, the fire in his eyes began to dim.

"How the hell can it be *fixed?*" His tone had sunk almost to its normal resonance.

"There's a lot of that ticking around," I told him. "Old pillows and stuff. Jen . . . One of the girls could cut a piece out and clean it and sew it on there so you'd never notice."

Curtis stared at the mattress as I spoke, undoubtedly envisioning busy fingers engaged in making it whole.

"How about the stuffing?"

"You can get that stuff in Farmington," I said, just as if I knew exactly where. Actually I had no idea that you could buy it at all.

Curtis pressed his lips together and nodded five or six times, like a man adding up costs in his head.

"You can take care of that?"

"Well, I can tell Kidder. I probably won't be here."

"Tell him to put it on my bill." He still had not taken his eyes from the mattress. He heaved a great sigh now, expelling the last of his anger, and shook his head in mild wonderment. "I just can't *stand* any kind of *slackness*. Especially after I . . ." He bit that sentence to a quick close, checking his temper. He looked up at Gordon.

"Well, let's get this thing put away for now."

Gordon hastened to grab one end of the mattress and together they fitted it back in the drawer, which the older man closed with great deliberation, driving it in the last fraction with a tap of his toe. He looked up at me, completely calm now, and nodded, wide-eyed.

"You won't forget now?"

"Not a chance."

Gordon offered me a gentle smile and I winked at him.

"I'll remind him," said Gordon.

"You do that," his father muttered, without the slightest conviction.

With everything in order finally, Kidder backed the horses out of the hovel and I helped him with the harnessing. We had little freight to carry, except Gordon's skis and the dressed-up luggage the two men had brought with them. I decided not to try to smuggle Jenny's rifle out on this trip, so I had only my own small kit to bring. Kidder tossed a pair of snowshoes in with the skis.

"Just for luck," he declared.

Only two people could easily sit up with the driver, so I found a place under one of the heavy blankets in back and hunched myself there against the cold. I had hardly settled when, to my amazement, the elder Curtis, the skirts of his great coat gathered high in one hand, clambered up over the tailboard to take a place beside me.

"You look pretty comfortable," he said. "Room for one more?"

I edged over quickly to provide extra room and threw the blanket back so he could get his legs underneath. I was still too astounded to do much more than mutter some mild agreement. Curtis took a long time to get his coat adjusted, the collar up to shield his face and the blanket snug about his knees. His body pressed close to mine. This intimacy with so majestic a figure robbed me of speech. Having been raised in a Boston neighborhood where "rich kids" were objects of scorn, as being pampered and sissified, and where their parents were usually thought of as the enemy, I had no resources for dealing on equal terms with such people. So I edged over as much as I could without robbing him of any blanket and vainly rifled the crannies of my mind for polite phrases that might fit this situation.

Then Kidder called out, "Everybody settled back there?" Mr. Curtis and I cried out together: "All right!" Kidder yelled, "C'mon, BILLY!" and we started off, the sleds bumping sturdily on the uneven road. Curtis and I were thrown together suddenly and he laughed almost in my face.

"Rough ride!" he exclaimed and I happily agreed.

"Gordon tells me you're a college man?" His tone made this into a question. Unwilling to elaborate any further on my falsehood, I merely told him: "I didn't finish."

"Couldn't make the grade?"

This suggestion really irritated me, as did any intimation that I was short of sense, and I responded sharply.

"No! I had good grades! It was just the money."

Gordon leaned over to look down on our heads.

"Good grades in what?"

His father rolled his eyes up at him.

"This is a private conversation," he said mildly.

"Excuse ME!" Gordon hunched his head down between his shoulders in make-believe humility, and turned forward again. His father spoke to me in a slightly lower tone.

"In what?"

"Well, everything. I was certified into Harvard." This was indeed a fact. But I had never had any urge to attend Harvard. Quite the opposite. Going to classes at Harvard and living at home would have been like continuing high school, somehow unmanly.

Mr. Curtis pursed his freezing lips into a vain effort to whistle.

"Straight A's then," he said and looked at me for confirmation.

"That's right." This was not a fact. But it seemed to me

that B plus in physics was close enough so that a very slight nudge would have qualified it. That was the only one I had missed.

"Damn shame you didn't finish. A man *needs* a college degree to get *anywhere* these days. And combined with the kind of practical experience *you're* getting." Mr. Curtis gave a twist of his head to indicate the impossibility of putting a sum to the combination. I began to feel slightly exhilarated, as if I had control of this exchange.

"I won the prize in Latin," said I. The moment I uttered that sentence I cringed at its fatuity. I imagined Kidder flinging it back at me as we worked a crosscut saw together. Mr. Curtis did chuckle slightly.

"I don't imagine that's much help to you here."

"Nope." Now I wished I had turned the talk into another pathway. I could sense some very tricky footing ahead. Mr. Curtis himself steered me to safety.

"Go in for any sports in school?"

That was a subject I had no immediate need to lie about.

"Football," I said confidently. "I ran a lot too."

"What distance?"

"Six hundred."

"That's an indoor distance. You run out-of-doors at all?"

"Well, no. I played baseball in the spring."

"That's pretty damn good. Wonder some college didn't come after you."

"I ran my own baseball club in the summer." This, a simple truth, was also my means of dodging any question about my making the school team. I always felt I should have, anyway. And I told myself that the school coach, after my club had licked his one afternoon, might have agreed with me.

"Well, say. You must have been pretty damn versatile! Light for football though, weren't you?"

"I guess so. I didn't get taken out of the play too often." This was a gross exaggeration. But I comforted myself with the fact that once, as a man "hard to take out," the coach had selected me for his blockers to work on. I knew full well, however, that this skill would cut very little ice among the monsters I had seen on college gridirons. At six feet and one hundred and thirty pounds I was not too far removed from the privilege of playing the "Human Skeleton" in a circus sideshow.

I thought I sensed a degree of skepticism in the manner in which Curtis eyed me now. I could feel the embarrassment creeping slowly up my neck and I fell quickly silent. But Curtis just shook his head slowly.

"Damn shame," he muttered.

We rode on then for two chilly miles or more without speaking. And I sat glumly hugging my knees, thinking, like the man in the song, "thoughts I could not utter."

Once Kidder looked back at us, perhaps concerned about our silence.

"Keeping cool back there?" he cried.

"Just fine," I croaked. Curtis straightened up to face front, pulling the blankets loose.

"How about you fellows? You all right, son?"

Gordon seemed to pull his head down more deeply between his shoulders.

"Great!" he replied. But there was a shiver in his tone.

"Pretty near there!" his father assured him. Then he turned to settle himself and to tuck some of the blanket about me again.

"Any plans for the summer?" he said, just as if our talk had not been interrupted.

I had to clear my throat vigorously so the voice would come out.

"Guess I'll go back to the hotel in April."

"Mooselookmeguntic?"

"No. Mountainview. It's over on Rangeley."

"Oh, yes. Very nice." He squinted at me as if he were trying to size me up anew.

"Ever think about going down to the city? If you don't go back to school?"

I sensed an offer of some sort hatching and I tried to gird myself to show proper interest.

"Maybe sometime. If I could get into newspaper work or something."

Mr. Curtis raised his brows in mild amazement.

"Oh. That's your line? Any experience?"

"Not really. I kind of fooled around." I was unable to invent a more robust falsehood than that. I *had* done two reports for the school paper.

"You'll need some more college for that, won't you?"

"I suppose so. Some guys get right into it."

"What about selling? You ever think about that?"

Dismay began to seep into my heart. *Selling*, for God's sake!

"Not much." I wondered if I dare ask him about a job here at the storehouse, with Kidder. But he was not to be interrupted.

"I've had this idea . . . that is, we've been planning for some time on opening an office in New York, selling our paper towel line direct to the consumer. More or less of an experiment. But I'm convinced it may go like a house afire, if we can get some fellows into it with a little old-fashioned get-up-and-go. Fellow like yourself. I mean, you don't seem to be one of these clock-watchers. You might get in on the ground floor there and really go places. It'd be good too to have someone on the ground there who at least knew *something* about where our product comes from."

312

I could think of hardly any responses that would not commit me to a dismal fate.

"Sounds good," I murmured.

Curtis straightened around and tucked himself tightly into the blanket.

"Well, think about it. If we decide to go ahead with this. Well, I guess we know where we can find you."

"Kidder knows anyway," I told him, trying to put some enthusiasm into my tone. But what the hell would I do if he should *really* make me the offer? (I need not have worried on that score, for I was fated never to see Mr. Curtis or Gordon again.)

After a few minutes of contemplative silence, Mr. Curtis spoke in a solemn tone.

"College ought to be first choice, though. If you can possibly make it."

"You're probably right," said I.

He was too, in a way, for I did go back to college, although I did not stay. But that was no part of my plans at this moment.

There was a long straight run on a well-traveled road before we came to the storehouse, and old Billy, with the sweet aroma of the stable in his nostrils, trotted along like a carriage horse, all the harness bells a-tinkle.

We — all but Kidder — disembarked at the office camp, a small red bungalow with a brick chimney from which gray smoke crawled to drift along the rooftop. The office door opened as we climbed, stiff-jointed, out of the wagon, and who should stand there but Stapleton, eyeing us in that speculative way he had, as if he just might start to smile.

"Howdy!" he called.

Mr. Curtis turned and frowned at Gordon.

"What's *that* galoot doing here?" he muttered. Gordon shrugged. There was no opportunity to answer.

"Come in and set!" said Stapleton. "Get the frost out of your bones."

We entered single file, each of us beating his hands together as if keeping time to a drum. The small black stove in the office dealt us a torrent of heat that engulfed us instantly. All the hats came off and we stamped and whuffed to expel the chill.

"Get that woman put away?"

Curtis addressed Stapleton as if he were asking an office boy if he had remembered to deliver a package.

Stapleton studied the man coolly for several seconds before he replied.

"All tucked away safe," he said finally, in a tone that was barely audible.

Gordon had shucked his coat and stood with his back to the stove, rubbing heat into his rear end. His father sought a hook to hang his own precious coat on, then joined Gordon and me as we performed the same ritual. Stapleton sat himself down and fixed his eyes on me. I met his glance for just a moment.

"What's up, Stape?" My voice actually squeaked a little.

Stapleton nodded his head a few times and kept his gaze on my face.

"What'd Kitty Dunham want to shoot *you* for?"

For some reason this question filled me with alarm. Was I suspected of some dark deed? Had Stapleton come all the way out here just to question me again?

"Jesus. I don't know. She was just crazy. She said I stole her brother's car."

"Her brother's dead."

I took this as a suggestion that I had made up the story.

"I'm just telling what she said. Kidder was right there."

"I don't doubt it. Half the time she don't know if she's

afoot or a-hossback. But she must have had it in for you all the same, for something. You ever poke around her place?"

"Jesus, no!" I had hardly formed the words when I recalled that indeed I had — not poked around exactly, but I had been there.

"Well, I was by there once. I went down that day I saw you in MacKenzie's. I was kind of curious, so I just thought I'd take a look."

"You and everybody else in town," said Stapleton. "Didn't see nothing special though?"

"No. Just a guy coming out of the shanty. And somebody in the window."

Stapleton nodded as if I had given the correct answers to a quiz.

"What'd he look like? This guy?"

"I never saw him before. He had kind of a hatchet face and needed a shave."

"Didn't say nothing to you?"

"Not a word. As a matter of fact, when he first saw me he started to duck back into the shanty. Then he came out again and walked into the back door of the house, not even looking at me. I thought maybe he forgot to button his fly."

"Maybe he did."

I was aware that old man Curtis had been glowering more deeply as this quiz progressed. Now he broke in sharply.

"What's this questioning about, sheriff? You have any complaint against this boy?"

Stapleton turned his clear, icy gaze on Curtis, and held it there for a good fifteen seconds before he replied.

"Just wanted a few answers. Just curiosity."

Curtis flushed at this. His nostrils widened as if he had been running.

315

"Well, when your curiosity is satisfied I suggest you put a few questions to the men who pulled the trigger. They must be two townships away by now. Or over the border."

Stapleton nodded, cool as ever, just as if this were a polite exchange.

"They may be," he said. "Wouldn't surprise me none. But I got a question for you, if you don't mind."

Curtis had to swallow hard. Gordon was staring at him, transfixed.

"What's that?"

"That linesman's shack up there where we found this lady." Stapleton made a vague gesture toward the west. "I figure that's on company land."

"Probably is. What about it?"

"Like your permission to burn it down."

Curtis could not have been more surprised if Stapleton had asked for his daughter's hand in marriage. He seemed to have lost most of his breath and his response came out in a half whisper.

"Burn it *down?* What the devil for?"

"It's a danger," said Stapleton. "Folks going up there like she did, to tie one on. No telling what might happen. Hell of a danger of fire."

Curtis could not deny this but he seemed reluctant to end this encounter peacefully. He looked at me and at Gordon, seeking inspiration of some sort. But we had no votes in the matter.

"I'll talk to my woods people," said Curtis. "If they've got no use for it, well . . . Kidder'll let you know."

He turned away abruptly then and put one hand on Gordon's arm.

"Come on, son, let's you and me go get into a hot shower. I imagine you could use one." He turned to me and extended a fleshy hand. "I'll say good-bye for now. Think about what I told you."

"I sure will," I told him, although I had done all the thinking about it that I intended.

"I'll be up in a minute," Gordon said. "I want to stay with Bob for a while."

Nothing was going to detain his father. He started at once for the door.

"Don't stay in this heat with those warm clothes too long. You'll catch the deuce of a cold."

Gordon grimaced at me.

"Yes, Father," he murmured. But his father was out the door, having exchanged no farewells with Stapleton. Gordon smiled at the sheriff.

"I'd sure like to be there when you burn that place down."

Stapleton nodded pleasantly.

"Be quite a show. We'll have to do it soon, before the snow goes. You're welcome to watch."

"I can't," said Gordon. He put one hand on my arm. "You going to be around awhile? You have to wait for Kidder anyway, don't you?"

I agreed that I did, and Gordon put out his hand.

"Well, in case I don't see you." He shook my hand fervently. "You going back with Kidder?"

"I think so. We got some stuff to clean up. Then maybe we'll go watch the fire."

"I wish the hell *I* could. But I have to be in class day after tomorrow. So why don't you take a run down and see me at Hanover someday? I'll show you around. Just ask for the Deke house. Or maybe, if you're going to be around this summer . . ."

"Oh, I will be. All summer. I'll see you sometime."

"Great! We'll take old Kidder out and get him drunk."

He turned then and told Stapleton, very formally, that he was pleased to have met him. He winked at me.

"Remember. The Deke house. Just bang on the door until they let you in."

"I will," said I. But I never did.

Stapleton motioned me to sit beside him. The office was a barren little room, with a desk not much bigger than the stove. There was a telephone on the wall and a large map of the area in a skinny frame right beside it. The floor was burnished until it gleamed like metal and our shoes had left puddles that refused to spread.

"What's happened to Kidder?" said Stapleton.

"He's stabling the horse."

"Taking his time about it. You going back up with him?"

"Yeh. We got some things to fix. Windows and stuff. And he wants to mend the phone line as soon as the snow goes down enough."

Stapleton nodded.

"If it comes off warm that snow'll be gone in a week. Except in where it's thick."

"I hope so." I offered this comment just as solemnly as if we had been appraising our hopes for life everlasting.

"That's why I want to touch that cabin off as soon as we can. Can't do no harm when there's snow on the ground."

"You going to clean all the stuff out?"

"Have to."

The door opened at this moment and Jenny came in, with Kidder right behind, grinning.

"Look what I found out loose in the yard!"

Jenny laughed at this, in her usual generous way, lifting her chin and shaking her head to indicate how useless it was to try to keep Kidder from making idiotic remarks.

"Have you been here all this time?" I asked her. "You didn't go home?"

318

Jenny looked straight into my eyes and gave me her wide-open smile.

"Too damn lonesome. So I thought I'd visit Cora for a spell. That's the thing about the winter. Gives you a chance to visit."

I know I flushed at what I took to be the hidden meaning of this remark. It was not my intent to devote the early spring to keeping Jenny company. But Jenny seemed not to notice, although she looked straight into my eyes all the while. I wondered if perhaps Kidder might urge her to come back with us. But before Kidder said a word, Stapleton pointed a rather grimy finger at Jenny.

"Just the young lady I want to see. I know you said one time back along that you knew where Kendall kept his liquor. Where was that?"

Jenny's face turned grim. In that day, you did not turn in your neighbor to the liquor sleuths, no matter how ill he might have used you. Stapleton merely chuckled.

"In the shanty, warn't it?"

Jenny's mouth dropped open.

"Well, if you knowed already, what'd you ask me for?"

"Just wanted to see if I guessed right. You see him getting liquor out of there?"

Jenny stared at Stapleton for a moment, her face still glum.

"Well, I didn't see anything I could swear to. I saw him carrying something in, something wrapped up in a newspaper. He give me a dirty look."

"A bottle?"

"Maybe. It looked like a bottle."

"Maybe a gun?"

This time everyone seemed struck dumb. Kidder spoke at last in a whisper.

"Holy Jesus! You think . . . Holy Christ!"

Stapleton shifted in his chair, as if he were preparing to deliver a speech.

"There's one funny thing about that business," he said. "Jack wasn't shot with no rifle. They found a forty-five slug right beside him. Made a hell of a hole coming out." He raised his eyebrows at Kidder. "You seen what one of them things can do?"

Kidder nodded. "Sure have."

Stapleton seemed to ruminate for a moment, then for some reason looked at me, as if he was about to bring me up to date on something the others knew.

"They found old Jack in his undershirt. Now no sane man's going out in this kind of weather in his undershirt!"

He seemed to be waiting for my comment, so I made a noise to indicate agreement.

"Course not," Stapleton went on. "What I figure is somebody done away with his shirt and coat, same way they done away with the gun. Other ways than that anybody could see he'd been shot up close. That make sense to you Kidder?"

Kidder allowed it did, and Stapleton, after a nod or two, took up the story.

"Well, if you wanted a place where nobody was likely to come poking around, I don't know any better place than the shanty. Couldn't be handier. Private too!" Stapleton snickered at this and we all dutifully joined him.

"Well, I don't know about the shirt and coat," said Kidder. "I don't know what kind of shape they'd be in after three or four months worth of shit been dumped on top of them. Maybe a little quicklime too. But the gun could be dug out, I suppose."

Stapleton looked at Kidder for several long breaths. His

mouth was clamped tight. He heaved himself into a new position and looked down at the toes of his boots.

"Oh, it'll be dug out someday maybe. Somebody might try it, but it won't be me."

"Oh, come on," said Kidder. "You must have shoveled shit in your day. Long's you've lived in this country."

"Shoveled plenty of it," said Stapleton. "But I ain't going to shovel this. I've found out all *I* want to know anyway." He made a grimace that bore a slight resemblance to a smile. "Like the big boss just said, I already satisfied *my* curiosity. Now somebody else can try it. If they can find anybody dumb enough."

Kidder looked truly alarmed.

"You ain't *quitting*, for Christ's sake?"

"No. I ain't quit. And I ain't a-going to. But I got word from — well, from somebody knows about such things — that they, I mean the head ones in Augusty, they're going to pick up my badge. Maybe next week."

Kidder got right out of his chair, as if he were seeking someone to fight.

"I'll . . . be . . . god . . . damned!" he declared. "That miserable old bastard! He got to Fessenden, didn't he? Them county folks jump when Jesus H. Fessenden farts. Wouldn't you know it, for Christ's sake?"

We all of us knew he was talking about Kendall. Fessenden, who owned about seventy-five percent of the vacant land in the township, was president of the bank and held mortgages on most of the homes. He once had a warden fired who dared "take" him for trolling live bait in a "fly-fishing-only" waters. He and Kendall, as the current phrase had it, were "asshole buddies."

Kidder and Jenny and I looked at each other in complete dismay. Stapleton gave a dry little laugh.

"Leastways," he said, "I'm going to have the satisfaction of burning down the bastard's hideout before I go. That's if I can get Kidder's permission."

Kidder, his face still flushed with anger, responded in a voice that shook the windows.

"You mean that old linesman's shack? Christ, yes! I'll touch the match to it myself! Whenever you say."

"Can't do it before morning."

Kidder looked to me for agreement.

"Right after breakfast?"

I was delighted at the prospect of striking such a spectacular blow for justice.

"Fine with me."

Jenny was watching us, white-faced.

"I'm going with you," she declared. "I ain't going to feel too safe at home alone."

"Come along, for Christ's sake!" Kidder shouted. "The more the merrier! We'll put on a Jesusly war dance." He grinned at Jenny. "You sure Bob won't give you no trouble? We can put a halter on him if you say the word."

Jenny was able to smile at this. She looked at me fondly.

"He won't give me no trouble. It's you and Stape I'm worried about."

Stapleton, for the first time since I'd known him, put his head back and released four or five barks of laughter, startling all three of us. Stapleton, for Christ's sake! I don't think any of us had ever known him to exhibit much more than a sort of wry snicker of joy over anything.

"I wish to hell you'd hang on a while longer, Stape," said Kidder. "I'd just like to see you stick it to that old lard ass. A stretch in Thomaston would do him nothing but good. Might even melt off some of the lard."

Stapleton shook his head vigorously.

"I don't believe Kendall done it," he said.

322

"You don't?" We all of us looked at him, mouths open.

"Hell, no!" Stapleton shook his head once or twice more. "Christ, I don't believe he could shoot himself in the foot if he tried for ten minutes. No. Kitty done it herself, surer than hell. I can't figure it no other way."

Jenny twisted her mouth in puzzlement.

"What would she want to *kill* the poor bastard for? She was getting all she wanted. Jack sure never interfered with the way she was carrying on with them two jokers. Or with anybody else, for that matter."

Stapleton nodded.

"I know," he said. "But Jack was like all them Dunhams. They's all the same way. Easygoing. But you can push one of them just so far, then they'll turn on you. I seen it time and again. Jack warn't no different. Probably boiling inside all the while. Then, when she rubbed it into him just once too often, it all come out at once and he fetched her a crack on the head." Stapleton banged his own head with the heel of his hand. "You seen how she took to wearing that Jesusly stocking cap even when she was in the shop. Well, when we picked her up in that cabin, I don't know if you noticed, but she had one hell of a welt on the side of her head. Her hair covered it and I suppose it faded a bit. But Jesus, it had more colors than granny's afghan. Purple, green, yellow, red. Had to be one hell of a sight when it was fresh. Must have used the butt end of a buggy whip. Most likely knocked her flatter'n a turd. Then when she come to, she went gunning for him. Just like her. And just like them other critters to cover up her tracks for her."

Jenny kept shaking her head.

"You mean she just up and shot him because he laid into her? After all these years?"

"Sure. He warn't no more than a half-assed monkey in her life. You own an animal that turns on you and you just

323

take it out in the woods and do away with it. Right? Well, Jack never even raised his voice to her for twenty years that I know about, then he went wild. Don't believe she wasted one minute thinking about it. Just got that old hoss pistol out and waited out there in the brush till he came back from Oquossoc. He was shot out there in the road, for Christ's sake, right after dark. She must have walked right up to him and let him have it. Probably never spoke a goddamn word. Just BANG! Hit *me*, will you, you miserable little bastard?" Stapleton, using his thumb and finger for a gun, jabbed the air in front of him. "Course, afterwards she had sense enough left to know she had to do something about covering up. So she . . . Well, you know what she was like as well as I do."

"Well, she was about the only one in the county crazy enough to do something like that," said Jenny. "But Jesus! What did she come after Bob and me for? What'd we ever do?"

"Don't doubt she seen you looking around down there and figured you was on to something. Or old single-tit Bailey heared you talking around and come running to tell her."

Jenny's eyes grew suddenly wide.

"I never 'talked around'!"

Stapleton chuckled.

"Well, Jesus, Jen, nobody has much trouble hearing you when you *do* talk."

"Oh, I suppose." Jenny managed a wry smile.

"Anyway," Stapleton went on, "she's the one that done it, surer'n hell. For one thing, I always knowed she had that Christly forty-five. Makes a noise like a cannon. And another thing, Bailey and Kendall warn't near the place when it happened. They was over to Rangeley, the two of them. A dozen people seen them there. So, unless you

324

believe that Frenchmen story, which I don't, you have to figure it was her. Only I couldn't figure what the hell she done it for until I see that welt she was wearing. She wouldn't have stood still a minute for that kind of stuff. By Jesus, the minute she could walk she'd a been after him. Her father was the same way. Fiery! Only he warn't nearly so crazy wild."

"I still don't see why the hell she picked me out," said I. "Where'd she get this idea about the car?"

"Who the hell knows? She warn't right in the head even when she was sober. It never took much to get her going. She was always dreaming up things people was trying to do to her. Like the time she wanted me to take that poor miserable Ed Veinot in for trying to climb in her window. Why the poor son of a bitch had all he could do to get through a *door* with that twisted leg of his. I tell you, it got so I'd head for the hovel when I seen her coming."

"So what happens now?" said Jenny. "The old bag going to get *away* with it?"

"Nothing happens, far as I'm concerned. They wouldn't do no worse to her if they *could* pin it on her. More'n likely they'd make out it was self-defense and she mightn't do no time at all. This way she's going to stay right where she is for a long, long time. As for them other two jokers, well, at least I put the fear of God into them. They'll be pissing blood for weeks, believe me."

We set off soon after dawn next·day, with Jenny and me together in the back of the wagon along with several lights of glass, six squares of shingles, points, putty, an assortment of tools, fresh beef, bacon, and God knows what else in burlap potato sacks. Jenny and I shared the big horse blanket and she made no bones about snuggling close.

We ate a cold lunch in the camp kitchen and hurried then up the long lake, with Stapleton in the lead, to hold our pagan celebration. There was crust enough to hold us without snowshoes and the day had grown almost warm, so that sweat ran on our faces.

We smashed open the cabin door with heathen delight and set about fervidly to toss all the contents, precious or not, far out on the snow, where we could sort them at leisure. Even Kitty's rifle went sailing to join the crockery, the tableware, the pots, the bedding, the boxes of shotgun shells, and the tin plates that skittered across the gleaming snow. Kidder took up the two table lamps, emptied out the kerosene into an empty can, and set the lamps far out on the snow with a modicum of care. There was old paper and kindling aplenty to make good tinder at all four corners and Kidder doused each bundle with kerosene. Then he and Stapleton, nursing matches alight, set every corner to burning. The flames rose slowly at first, fading and fluttering, then found fresh nourishment and began to jump from board to board, until they gathered at last into one roaring tower that forced us all to hop away from the heat. Stapleton, squint-eyed and grim, seemed to count every purlin that burned through and toppled into the bottom of the blaze. He and Kidder from time to time would move close, faces averted, and kick some partly burned remnant back into the flames.

The fire burned on for more than an hour, scorching the nearby spruces and crumbling green needles on the top branches into glowing black cinders. We all stood there until the fire had subsided to flaring coals, reading our private intentions in the flames. Then we picked about among the scattered furnishings to find whatever was worth keeping and put them into small piles to be retrieved another time.

"Anything you leave behind in an abandoned cabin belongs to the landowner," Kidder announced.

"Kidder's law," said Jenny. Kidder laughed.

"That's the only law counts for a damn hereabouts right now."

That scene of the four of us happily moving about the ravaged and smoking cabin, like kids around a bonfire, is the one that remained with me longest. Stapleton, scorning food, drink, or transportation, soon took off on his snowshoes on the ten-mile hike to home. Kidder and Jenny and I repaired to camp and completed our chores there. I believe Kidder took Jenny out first, for her job at the laundry was beginning, while Kidder and I stayed together there with not much to do but consume the groceries, patch a roof and a window, take the boards down from the windows of private camps, and count the homebound geese that flew overhead in multitudes, with a noise like distant dogs barking.

In the two seasons still ahead of me in that country, I saw Jenny but seldom, for she moved to California with her sister the following fall. When I saw her again in Oquossoc, forty years later, she had turned into an old woman, but a lively and happy one, and she greeted me with all her former heartiness, this time leaping right into my arms and hailing me, correctly this time, as "Old Man Smith!"

Kidder and I met next some twenty-five years earlier than that, when he was in charge of all wilderness operations for the company. We remained constant hunting and fishing companions then for as long as he lived.

As for Stapleton, I saw him from time to time during the next two years, but always at a distance. Yet I have held in my heart to this day the words he uttered that afternoon, as he shook my hand.

"You going to stick around this country, young feller?" said he, with that half smile on his face.

"I don't know. I'll get into something different one of these days. I may go down to New York."

I really had no such intention then, and was probably just reacting to the hint by old man Curtis that I might have a future there. But Stapleton responded with a nearly complete smile and beaming eyes.

"New York! Good! Good! They need fellers like you down there! They need a lot of them!"

And, as my history developed, I did indeed go down to New York before many more years had passed, and found no evidence there that anyone was conscious of the fact that I was needed. Still, I have relished that unearned and undeserved accolade ever since.